Transformation, Co-operation, and Conversion

NATO ASI Series

Advanced Science Institutes Series

A Series presenting the results of activities sponsored by the NATO Science Committee, which aims at the dissemination of advanced scientific and technological knowledge, with a view to strengthening links between scientific communities.

The Series is published by an international board of publishers in conjunction with the NATO Scientific Affairs Division

A Life Sciences **B Physics**	Plenum Publishing Corporation London and New York
C Mathematical and Physical Sciences **D Behavioural and Social Sciences** **E Applied Sciences**	Kluwer Academic Publishers Dordrecht, Boston and London
F Computer and Systems Sciences **G Ecological Sciences** **H Cell Biology** **I Global Environmental Change**	Springer-Verlag Berlin, Heidelberg, New York, London, Paris and Tokyo

PARTNERSHIP SUB-SERIES

1. Disarmament Technologies	Kluwer Academic Publishers
2. Environment	Springer-Verlag / Kluwer Academic Publishers
3. High Technology	Kluwer Academic Publishers
4. Science and Technology Policy	Kluwer Academic Publishers
5. Computer Networking	Kluwer Academic Publishers

The Partnership Sub-Series incorporates activities undertaken in collaboration with NATO's Cooperation Partners, the countries of the CIS and Central and Eastern Europe, in Priority Areas of concern to those countries.

NATO-PCO-DATA BASE

The electronic index to the NATO ASI Series provides full bibliographical references (with keywords and/or abstracts) to more than 50000 contributions from international scientists published in all sections of the NATO ASI Series.
Access to the NATO-PCO-DATA BASE is possible in two ways:

– via online FILE 128 (NATO-PCO-DATA BASE) hosted by ESRIN, Via Galileo Galilei, I-00044 Frascati, Italy.

– via CD-ROM "NATO-PCO-DATA BASE" with user-friendly retrieval software in English, French and German (© WTV GmbH and DATAWARE Technologies Inc. 1989).

The CD-ROM can be ordered through any member of the Board of Publishers or through NATO-PCO, Overijse, Belgium.

Series 4: Science and Technology Policy – Vol. 7

Transformation, Co-operation, and Conversion

edited by

Kazimiera Prunskienė
Lithuanian European Institute,
Vilnius, Lithuania

and

Elmar Altvater
Free University,
Berlin, Germany

Kluwer Academic Publishers

Dordrecht / Boston / London

Published in cooperation with NATO Scientific Affairs Division

Proceedings of the NATO Advanced Research Workshop on
Scientific and Technical Co-operation of the Baltic States in the
New Europe and the Conversion of their Industry
Vilnius, Lithuania
June 5–8, 1995

A C.I.P. Catalogue record for this book is available from the Library of Congress.

ISBN 0-7923-4178-3

Published by Kluwer Academic Publishers,
P.O. Box 17, 3300 AA Dordrecht, The Netherlands.

Kluwer Academic Publishers incorporates the publishing programmes of
D. Reidel, Martinus Nijhoff, Dr W. Junk and MTP Press.

Sold and distributed in the U.S.A. and Canada
by Kluwer Academic Publishers,
101 Philip Drive, Norwell, MA 02061, U.S.A.

In all other countries, sold and distributed
by Kluwer Academic Publishers Group,
P.O. Box 322, 3300 AH Dordrecht, The Netherlands.

Printed on acid-free paper

Printed in the Netherlands

CONTENTS

LIST OF CONTRIBUTORS

KAZIMIERA PRUNSKIENĖ
Former Prime Minister of the Republic of Lithuania
President of The Lithuanian-European Institute
Vilniaus St. 45-13, 2001 Vilnius, Lithuania

DR. ADOLFAS ŠLEŽEVIČIUS
Prime Minister of the Republic of Lithuania.

PROF. DR. HABIL. V. DOMARKAS
Minister of Science and Education of the Republic of Lithuania
A.Volano 2/7, 2691 Vilnius, Lithuania

DALIA BARDAUSKIENĖ
Planning Division Head at the City Department of Architecture and
Building Co-ordinator of the Canadian Urban Institute, Vilnius Programme.

PROF. DR.W.M. GRUDZEWSKI AND PROF. DR. I. HEJDUK
The Institute of Industrial Organisation and Management
Industry "Orgmasz"
87, Zelazna st. 00-879 Warsaw, Poland

DOC. DR. KATARZYNA ZUKROWSKA
Professor at the Institute of Development and Strategic Studies
Plac Trzech Krzyzy 5, 00 507 Warsaw, Poland

DR. .MIKHAIL MYASNIKOVICH
Deputy Prime Minister of the Republic of Belarus
Candidate of Economic Sciences
Dom Provitelstvo, Minsk, Belarus

LEONID F: ZAIKO
Director of the Institute of Independent Research, NCSI "East - West"
Minsk, Republic of Belarus

V.D. NOZDRACHEV
President, Jupiter Stock Company
Krestovskij Pr.9, P.O. Box 934,
197042 Saint Petersburg, Russia

PROF. B. SAUSKAN DR. C. OVERKO
The Kaliningrad Region Embassy
Juozapavičiaus 11, Vilnius, Lithuania

PROF. P.J. TEUNISSEN
Sociology of International Relations and European Integration
Groningen University, Holland

DR. MARTIN GRUNDMANN
Senior Researcher, Schleswig-Holstein Institute for Peace Research at the University of Kiel
Kaiserstr. 2, D-24143 Kiel, Germany

PROF.DR. ULRICH ALBRECHT
Free University of Berlin, Institute of International Politics and Regional Studies
Kiebitzweg 3, D-14195 Berlin, Germany

PROF. DR. ELMAR ALTVATER
Free University of Berlin, Department of Political Science
Ihnestr. 21, D-14195 Berlin, Germany

DIPL.-ING. DIETER HANEL
MaK System Gesellschaft GmbH
D-24159 Kiel, Germany

PREFACE

The scientific-technical potential of the Baltic States and their possibilities for co-operation with Europe require in-depth, specific analysis. With the deterioration of the structures of science and technology of the former Soviet Union, the severance of communications with former clients, industry and science and technology, the Baltic States - the leaders of the former Soviet Union in this area, faced a difficult new situation. The government budgets of these countries, due to economic decline, are not capable of financing scientific research and project studies, and industry has lost a large part (in some branches this loss reaches even 2/3 or more) of its potential, losing both funds and interest in the sciences.

The conversion of industry in the Baltic States is tied with the whole of its restructurisation, which still has not attained more precise directions for a new specialization or connections to the international market. The earlier dominant branches, such as machine production, electro-technology, radio-technology and the electronics industry, require essential modernization, which is possible only through co-operation with other developed countries, especially with the EU and NATO countries. This co-operation could include new mobilized capacities of science and technology. A longer period of stagnation and separation may adversely affect these capacities and lead to their dilution and weakening, due to the "brain drain" of more qualified scientists and specialists to the commercial sector, which does not require high intellectual levels.

The inherited sufficient basic specialist qualification and the acquisition of Western know-how, which, under new conditions, is oriented towards the new practical requirements of industry and infrastructure, bears significance for the scientific-technical and economic revitalization of the Baltic States. And this in its turn determines regional and European security.

The stimulation of this process is one of the more pertinent tasks for both national governments and international structures alike. In expanding the Partnership in the Name of Peace Programme through realistic work, the NATO Scientific Programme seminar, organized together with the Lithuanian-European Institute, received widespread interest in the region and had a certain influence on new analyses of co-operation possibilities and directions in industry, as well as on the securing of new ties. The participation of representatives from the scientific and practical spheres with their ideas (a majority of which are included in this publication), shows that interest in co-operation exists and may be viewed as a catalyst for the activation of this process.

The organizers of the seminar and the editors of this publication would like to extend sincere thanks to all speakers and authors, who agreed to publish their papers in this book. Special thanks to Dr. A. Jubier, of the NATO Scientific and Environmental Affairs Division, for his initiative and support regarding the seminar and the publication of materials.

Prof. Dr. Kazimiera Prunskienė Prof. Dr. Elmar Altvater

INTRODUCTION

The Scientific and Technical Co-Operation of the Baltic States in the New Europe and the Conversion of their Industry

NATO SCIENTIFIC PROGRAM
JUNE 5-8, 1995, VILNIUS

On June 5-8, 1995, the NATO Scientific Program and the Lithuanian-European Institute organised a seminar attended by over 140 participants from twelve countries: Lithuania, Latvia, Estonia, Germany, England, Belgium, Holland, USA, Poland, Russia, Belarus and the Ukraine. Opening the conference was a speech by LR President Dr. A. Brazauskas, followed by speeches by LR Prime Minister Dr. A. Šleževičius, Minister of Commerce and Trade K. Klimašauskas, Minister of Education and Science Prof.Dr.habil. V. Domarkas, LEI and IF president Prof.Dr.habil. K. Prunskienė, Dr. M. Myasnikovich - Deputy Prime Minister of Belarus, Economic Chancellor M. Leivo of the Estonian Republic, NATO Scientific Program representative Dr. Alain H. Jubier, Vice Dean of the Department for Political Sciences of the Berlin Free University Prof.Dr. E. Altvater, World Bank representative L. Jeurling, International Development Group representative B. Bishop, Prof. J. Frackowiak (Poland) of the Scientific Research Committee, Doc.Dr. K.Zukrovska (Poland) of the Institute for Development and Strategic Studies, Director of the Centre of Baltic-Nordic History and Political Studies Dr. J. Prikulis (Latvia), Kaliningrad Administration representative to Lithuania S.M. Overko, Prof.Dr. A.N. Bykov of the International Institute of Economic and Political Research (Moscow), Acad. A.N. Shlepakov, President of the Ukrainian Association for UN Support, V.I. Olshevskij from the Ukrainian National Science Academy, as well as a number of other individuals.

Participants ranged from scientists and business people - their diverse backgrounds generating much "brainstorming" for solutions to current scientific, economic and social issues, of which the following requisites were especially accentuated:

- to preserve, strengthen and effectively employ the scientific and technical potential of the Baltic region;
- to end the commercial crisis; to establish, through rational reform and a foundation for conversion, possibilities for the encouragement of new developments, of small and medium-sized businesses and of diversified economic structures;

K. Prunskienė and E. Altvater (eds.), Transformation, Co-operation, and Conversion, 1-8.

- to institute new co-operational know-how exchanges in Europe and a basis for partnership in the commercial sector; to acquire joint investment projects and to fulfill international market requirements;
- to strengthen the social and political stability of the Baltic region as well as security in Europe through the help of science and commerce.

Participants raised and analysed many-sided scientific-technical and commerce issues emerging after the fall of the USSR and its market collapse. Pessimistic evaluations included criticism of reform error, unfavourable conditions for investment and the incapacity of economic subjects (Prof. B. Melnikas - Lithuania, Prof.Dr. A.N. Bykov - Moscow). On the positive side was the vision of many possibilities not yet harnessed (L. Jeurling - USA, B. Bishop - USA and others). An energetic and substantial discussion ensued, concluding with formulated arguments, conclusions and suggestions.

NATO Scientific Program representative A. Jubier presented the programme itinerary, along with a description of its new territorial borders and orientation towards civil scientific research. These priorities, as well as the possibility for regional states to participate in priority research projects, sparked much interest among participants and mass media representatives, most of whom were not aware of the existence of the NATO Scientific Program or the involvement of and possibilities for divisional and regional scientists as well as commercial firms in its establishment. In this respect, the seminar was not only informative but also altered many existing views towards NATO itself and the content of its partnerships.

1. Scientific and Commercial Priorities in the Regional States

Each country has its own set of priorities, formed under specific environmental, geopolitical and historical conditions. New situations, emerging in an open market economy, introduce important changes in the priority sector. The lack of personal financial sources during the transformational period limits the utilisation of government means for the encouragement of priority sectors.

The Lithuanian Minister of Education and Science, V. Domarkas, and his colleague K. Klimašauskas, LR Minister of Commerce and Trade, stressed the possibilities for science and industry in the international market in the following areas:

Science: ecology, biomedicine, biotechnology, informational and communications technology, energy, agricultural sciences and food technology, metrology, surveying and quality control, various areas of economics.

Industry: machinery and equipment production, radio television and information equipment production, production of chemical materials and products, wood processing as well as cellulose and paper industry, garment manufacturing, fur, textiles and knitwear production.

In listing the most important areas for the potential of Ukrainian science and technology, Dr. A. Juzefovich mentioned the areas of materials research, cosmic and aviation technology, vessel construction, the creation of biotechnology and chemical technology, new energy resources and current information technologies, simultaneously stressing new management and organisational structures, along with the establishment of international co-operational and scientific-technical, financial and commercial groups. Vice Prime Minister Dr. M. Myasnikovich of Belarus and other speakers from this country listed the following priorities corresponding with BR specifics:

- in the scientific-technical activity sphere: the preparation of a foundation for the formation of economic, scientific-technical and social policy; public health and environmental protection; the development of energy and transport, information, telecommunications and relations; research of new resource-saving materials (including polymers) and technologies; laser physics; optics; microelectronics; radio physics; plasma and solid body physics; chemistry and mechanics;
- in the conversion sector: the creation of modern electronic elements; hardware technology and automated systems; telecommunications and other means of communication, home electric appliances and medicine technology; agricultural products processing technology and technically complex goods;

The priorities of the Kaliningrad region, as Russian territory entering into the Baltic region, were mentioned by S.M. Overko and Prof. V. Sauskan: transport, forestry, the agrarian sector, machinery production, fuel, construction complexes, the social-recreational sphere. It is thought that four technology parks or poli could be established here - for metal processing, ecologically pure technologies, ocean activity development and social-economic development.

Participants were quite relieved and interested to learn of the decrease in military potential in this region and experience in establishing economic zones.

Guests from Russia, Belarus and the Ukraine stressed the problem of conversion as related to the preservation of military potential, more strongly than Baltic participants. Their views ranged from political to ideological differences, including the direction of integration and the directional priorities of the West or East. Further discussion topics were the pertinence of many-sided relations, the role of openness to Europe and the international market, the activation of the internal market, the strengthening of industrial potential, obsolescence in the production sector during the last few years, compared with the leading trade sector and its reversal. The creation of jobs was called for, as well as the need to increase the standard of living and to realise social stability.

A seminar theme which was discussed and refuted was the pre-establishment of the contention to join NATO and of industrial conversion. The LR Prime Minister emphasised that orientation towards NATO does not mean more Lithuanian arms and priority for the defence industry.

In evaluating conversion as a decrease in military potential, Secretary to the Russian Embassy in Lithuania Dr.habil. V. Micheev touched upon its negative influence on the state of the fundamental and specific sciences, underscoring that these losses may be compensated only through the success of economic conversion. The Russian representative argued for NATO assistance, while preserving Russian scientific potential as the essence of a humanitarian country. Regarding this issue V.I.Olševskij (the Ukraine) noted that uncontrolled expansion of technology could very well have dangerous consequences.

With the experience of Poland in mind, Doc.Dr. K. Zukrovska argued that having correctly chosen the military industrial complex (MIC) conversion and commercialisation model, the defence potential of a country may be preserved as well.

More than once, the speeches of participants gave suggestions or encouragement for NATO to take steps in avoiding ecological disaster and to increase demands for science and scientific production.

Belarussian Vice Prime Minister Dr. M. Myasnikovich recognised the importance of industrial conversion in infrastructure and the area of scientific research, while decreasing the threat of war and encouraging economic development: He also drew attention to the specific situation in every country and external influences. In his view, conversion policy should be sought along with national military- economic policy; that is, there is a need for local arms capacity, export of military production, the stable functioning of the military industry and the rational utilisation of its financial resources.

The scientific and technical priorities for regional countries are not yet clearly formed, and much influence on their rational development comes not only from the international market but from effective strategy and mechanisms for influence.

2. The Restructuring of Industry and the Formation of New Markets in the Baltic Region

As was admitted by the seminar attendees, the conversion of industry in the Baltic region is taking place during a general economic re-structurisation. What is currently left of industrial capacity is no longer fit for military purposes, and new civil-oriented possibilities are neither conceptually or strategically based nor practically utilised. The most important role in this is played by the quality of existing cadre potential, which maybe long lost if the disintegration process of institutes and firms is not reoriented toward the creation of new, vivacious subjects, bearing active marketing teams and the ability to attract orders as well as co-operate with Western partners. An adequate national policy and realistic government assistance, however, are needed for this to take place.

Government agencies must gather and supply information regarding new technologies and innovations and their adaptation under given conditions.

Governments must create laws, a judicial base and institutional order. Without these a private market cannot function. World Bank representative L. Jeurling stressed the importance of an active role by governments in this area, critically evaluating the real situation in regional states.

The economic crisis, the large decline in scientific and industrial potential, is related not only to the inefficiency of the former Soviet system and the technological obsolescence of industry inherited from it, but there is a lack of conceptualism and strategy for currently undergoing reforms as well. Structural economic changes experience barriers posed by insufficient privatisation methods, lack of investment activity, lack of experience of reform specialists and governmental structures with their often apparent populist orientations. Government-implemented policy caters to self-interest groups and does not defend long term strategic goals to bring forth stability to its economic and social spheres.

Another concern highlighted by the speakers was the need for conversion to be based on market economic principles, creating a sufficient legal base and organisational systems. Doc.Dr. K. Zukrovska (Poland) mentioned the following basic problems:

- the creation of regulatory property laws;
- mechanisms of technology transfer;
- orders for military-industrial complexes (MIC) and a system for their control;
- preparation of an organisational and structural model for military industry.

On the practical side, the following items were outlined:

- combination of participation in MIC firm privatisation by government firm workers and external stockholders, while preserving a 51 % package of government stocks;
- the importance of international co-operation;
- the organisation of military-industrial holdings according to related production;
- the establishment of free economic zones in places of dual technology (i.e. Melece in Poland, Šiauliai Zoknių airport in Lithuania, etc.).

Acad. A.N. Shiepakov spoke of the need to take advantage of the experience of the US, which has a national technology transfer centre and other organisations, by which high technology is transferred to the private sector. Analogous systems are lacking in reform countries.

More unique know-how decisions by those of the former USSR and other socialist states combined with Western marketing, production certification and co-operation of Western, Central and Eastern European commercial producers, would spark a new development impulse, reopening benefits for all sides. The most favourable conditions of such a co-operation would be granted to those Baltic states located between Western Europe and Russia, as well as the Kaliningrad region. The latter, in the words of its representative S.M. Overko, sees its perspective as an open economic zone, possessing the status of independence and closely co-operating with neighbouring countries in the region.

Two once very large industrial firms received visitors from the seminar - the micro-production monopolist "Vilniaus Venta" and electro-technology firm "Elfa". Both firms experienced large crises, with a large part of their production terminated. Conversion strategies are practically non-existent, and work takes place according to sporadic orders; perspectives are hazy and certainly not optimistic.

These companies expect the interference of Western experts and clients, as well as realistic assistance and co-operational projects. Their potential is, in part, still preserved and may be once again easily consolidated. Without drastic basic changes, however, this potential may be lost in as little as 0.5-1 year.

3. The Preservation of Scientific Potential and Co-operation in the International Know-how Market

One of the greatest losses through the transformation of the economic system and disintegration of former scientific-technical ties among the Baltic states and the CIS countries, is that of the scientific collective which concentrated on fundamental and specific research and high technology. Many of the speakers, in stressing the problem of the devaluation of science, provided evidence of the sharp decrease in salary and real standard of living among scientists. An obvious flow of intellectual potential from the scientific, technical and industrial spheres to the commercial ("quick" business) area is felt by these countries. Because of shortcomings in protection of legal know-how (especially of patent rights), the markedly low salary of scientists, government financing difficulties and the state of crisis in industry, the best scientists and specialists in these countries - are practically "bought into" developed countries and, what is more threatening - into the South-Eastern Asian and Arab world, where this potential is used for criminal or war purposes. However, it is almost impossible to work with individual know-how leaders, their ideas and technical documentation, to take over an entire complex of conditions and foster the development of know-how (these are most often informal "generators" of ideas, who worked under science bureaucrats and remain in the shadows). Thus, as asserted by conference participant from St.Petersburg V. Nozdrachev as well as other speakers, methods must be sought for the revitalisation of regional scientific potential, incorporating the following factors:

- investment capital - for medium-sized business investments (including from the CIS) in know-how in the Baltic states;
- local scientific potential with its external ties;
- favourable conditions for investments and the know-how market in the Baltic States, where stability and its perspectives are much more trustworthy than in Russia and other CIS states.

Seminar participants mentioned an array of technologies in which former USSR science representatives are know-how leaders.

With the victory of open market economics ideology before the counter-position of the cold war and two militaristic blocs, the role of science in principle changes as well. In discussing this topic, Prof.Dr. A.N. Bykov (Moscow), stressed the international role of economic development in the globalisation process and in establishing opportunities for social and economic advancement, the possibilities for states in the international market, their competitiveness and place "under the sun" as well as participation in world processes.

Many reform states, including Russia, in essence changed their earlier activity zone (instead of the former USSR and CMEA), declaring integration into the world economy. The progression of practical activity contains many aspects, but new opportunities are unfortunately not utilised enough. The commercial production markets in these countries are slowly being taken over by modern Western countries (first of all the USA).

With the decrease in the need for science and technology, a large part of the research and concrete scientific research institutes, from the side of local military complexes and with the practically non-implemented conversion projects, lose their commissions and financing.

The Baltic states serve as the centre of Russia, the Ukraine and Belarus. The scientific-technical potential of Poland allows for the realistic assessment, preparation and fulfilment of scientific and economic area development and production projects. A large role could be played here by NATO and EU structures.

Speakers directed attention to the aspect of territorial economic development, including territorial planning, city conversion (the civil acquisition of demilitarising zones) and other actual topics in the NATO scientific program research and events.

Organising the consolidation of scientific technical potential and modern know-how and its supply in the international market, using modern marketing methods and the commercialisation of science (the latter was especially accented by World Bank representative L. Jeurling), is one of the most important factors for the development of science and revitalisation of industrial potential.

The participants from the Ukraine, Russia, Lithuania, Latvia, Poland and other countries accented the need for multi-lateral co-operation, putting into practise concrete scientific-technical programmes and investment projects for their realisation.

In admitting that in this process utilisation of Western know-how is a must, as well as co-operating with their know-how, industrial and financial structures, the priority of increasing partner-like relations was stressed. The suggestions of LEI and IF president K. Prunskienė, regarding a strategy for correlation of research, project study and investment project preparations through PHARE, TACIS, TEMPUS and other programmes, were supported by a number of seminar speakers. Among them were: BR Vice Prime Minister Dr. M.Myasnikovich, representatives from Poland Prof. J.Frąckowiak, Prof.Dr. I.Hejduk, Prof.Dr.Eng. W.M.Grudzewski, Ukrainian speakers and others.

In the view of seminar participants, it is mandatory to seek equal co-operation in the business sense as well as in the financial sense. The one-sided work completed by Western experts, without including the strongest local abilities and their non-financing, does not form possibilities to co-ordinate work results, taking into account local conditions. The discrimination of local experts pushes out the most capable of them to more profitable commercial spheres, although certain participants (among them B. Bishop - USA), however, noticed that this is not an absolute loss, since competition is necessary also in the business sphere.

When transferring from the project to the investment stage, the possibility of permanent work is lost, as well as for the accumulation of experience, necessary on all levels of management.

The problem of worthy co-operation of local experts and specialists with Western colleagues is not only their own problem. This involves government consolidated efforts, in defending this position towards international structures.

Seminar participants await assistance from NATO scientific structures, stressing that the re-installment of scientific-technical potential and its involvement in the international know-how market would aid regional industry, infrastructure and their economies. This is one of the most important factors of social and political stability and international security.

Technology parks and poli as well as international co-operation in their establishment were also discussed, with the suggestion raised to establish an international co-ordinating council.

With the conclusion of the seminar, a few agreements between participating scientific-research organisations were signed, regarding long term co-operation, the exchange of specialists and organisation of joint activities.

MECHANISM FOR SCIENTIFIC-TECHNICAL AND INDUSTRIAL CO-OPERATION BETWEEN CENTRAL, EASTERN AND WESTERN EUROPE

KAZIMIERA PRUNSKIENĖ

The character of international scientific-technical (ST) co-operation, its maturity and productivity and the application of its results, in the most general methodical sense, are determined by:

1. the aims of co-operation; the interests of its participants/partners;
2. the potential of the partners, its adequacy for both common goals and the potential of other partners (partner maturity). This is very important in moving from ST assistance to a working partnership - from the provider of assistance and recipient to the role of partners and real partnership status;
3. external and internal preconditions for co-operation, favourable general context; political stability of countries, the activeness of their foreign policy and international relations, economic model rationalisation, the character of geographical, geopolitical as well as historical and cultural traditional factors;
4. specific mechanisms: legal, organisational-institutional, economic-financial, their functioning and flexibility.

The processes for political and economic transformation, which are taking place in Central and Eastern Europe, including the Baltic States, in essence changed the possibilities for international co-operation and its character. Above all, this is apparent in:

- overcoming the former forced isolation from the Western world and one-sided co-operation within the boundaries of the former Soviet Union, openness toward Europe and the world market, the appearance of new possibilities in fulfilling the functions - a "bridge" - between East and West, and their gradual acquisition;
- broad diplomatic inter-state relations and the establishment of international agreements. June 12 marks the signing of an agreement between the Baltic States and the EU (Association);
- The mechanisms of a market economy are beginning to take shape and function: economic interests become the driving force in developing and utilising ST potential.

It can also be constituted that international co-operation is taking place most actively and is expanding most rapidly in the trade sector. Thanks to a more active trade of goods, our formerly sparse internal market, with its supply of goods and state provisions, has begun to approach Western levels. This sphere has

K. Prunskienė and E. Altvater (eds.), Transformation, Co-operation, and Conversion, 9-12.

attracted a large number of active industrial and scientific members, entailing negative results as well for the maintenance of ST potential. In other words, integration in the goods market in some areas is moving ahead more rapidly and effectively than in the ST and industrial spheres. Looking at the foreign trade balance, already approximately 50% of the total Baltic import-export balance is with the Western market. The character, however, of investment processes in industry, industrial co-operation with Western partners in the area of high technology and co-operation and competition in the know-how market is rather fragmentary. A large part of industrial and ST potential is struggling not only because of the lack of perspective or its inability to reach world levels, but also due to a number of other reasons, which are not easily solved and can only be solved one step at a time.

I will attempt to schematically assess a few (in my view realistic) problems connected with the expansion of co-operation and the establishment of effective mechanisms for co-operation.

1) One problem characteristic of the times is the devaluation of ST potential of new states in the international market to less than its real value - the difficulties of realising the engineering ideas and intellectual products in the newly attained international market to corresponding deficiency areas of infrastructure: project management, more mature consulting, etc. and the difficulties in establishing new subjects in the general context of economic crisis.
2) Internal communications are in disorder on a national level between the ST sphere and industrial structures. Large industrial firms have almost ceased placing scientific and technical orders and their co-operation is fragmented. Industry, which only functions to 1/3 of its capacity in Lithuania, is no longer capable of financing and supporting new research. Budgetary financing for science is also very limited.
3) The experience of scientific and technical structures in the know-how market is insufficient, with their lack of confidence and inflexibility to corresponding rules and conditions of competition. The problem of competitiveness in the broad sense ranges from mentality to technology, management and up-dating of information and marketing.
4) The complexity of ST of know-how marketing and the degree of its management, which is much smaller than in raw materials trade, are still insufficient.
5) The EU programmes are a significant form of ST co-operation: PHARE and its programmes ACE, JOPP and TACIS. Various economic areas and stages of innovative processes are covered within the frames of these programmes, from the conception of ideas and their preparation and project research to the financing of investment projects. It is still a long way to mature partnership and co-operation between East and West. This in part is apparent in local partners - in the inadequate activeness and parity of both government structures and in the spheres of science and business. The lack of precedent for initiative poses a problem. Parity is lacking in tender conditions, where local partners receive a

second-hand, not wholly valued role. In financial decisions, e.g. division of funds, local partners are squeezed out to 5-7%.

6) It is mandatory to strengthen the responsibility of governments and efforts in striving to utilise ST potential through:
    - the appropriate utilisation of budget funds in priority ST areas;
    - determination and initiative in offering suggestions in the EU and other structures regarding equal conditions of co-operation in PHARE and other programmes;
    - consolidation of regional government activities in this area, altering the traditions of co-operation between East and West;
    - more information for local experts and business structures on new projects.
7) Financing is not of a systematic character. Many significant branches of science have lost it altogether because of budgetary difficulties, without receiving new sources.
8) The utilisation of ST potential and its development without the need for investments, without active investments in the branches of the economy, is impossible.
9) The transition from research to the investment realisation phase is difficult. One reason for this is that research is often conducted without appropriate responsibility. Thus the problem of conducted research study application and its continuation is a realistic problem. If work is conducted only by foreign experts, the processes are not developed for their local application.
10) The activeness of ST potential may and should be pronounced through the formation of project supply. Lithuanian government structures have prepared such suggestions as well as structures in Russia (St. Petersburg). Unfortunately, their success among investors is small. Perhaps this is due to their unilateral development. The market often does not acknowledge these local priorities and instead dictates its own conditions, which are sometimes completely unexpected.
11) The role played by the depth of privatisation - Western partners must work together with real local partners, especially when speaking not of new objects but of the modernisation of former industrial giants. When a firm does not have a real owner but thousands of small stockholders, such a situation does not satisfy them. Investments are not attracted.
12) The specifics of practical models for co-operation in the current stage of integration: the international market is inclined to first trust the production of more simple products. Organisation of more complex technical goods (automobiles, devices, electronics) requires rather a complex mechanism of preparation as well as the consolidation of participants, i.e. manufacturers of ships, cranes or machine-tools in the West who are interested in transferring a part of the manufacturing construction to us. Having conducted research, it becomes apparent that technological problems have not been solved: welding, painting, etc., and the certification ISO-9000 assessment. Sometimes the clients

will complete this themselves but with exclusive rights this has little perspective. Having a long-term order, it is possible to complete necessary certification of the conditions of transport, logistics, and financing and their correlation.

Often transactions do not take place because no decision for payment is made. It is a typical situation when the Western purchaser agrees to pay for goods only under the conditions of the CIF - Western European Port or at best FOB, one of the Baltic ports (i.e. FOB - Klaipėda Port). And the Russian or Ukrainian product manufacturers, due to regulations, cannot export without pre-payment. Transport organisations also require pre-payment, which, under FOB or CIF conditions, is financially problematic for the Eastern partner. Thus the following is a typical situation in such a market: the client cancels the order for a necessary good at a reasonable cost, not having found suitable means of payment for all parties, and half of the production capacity is sold, regardless of food being demanded in the market. Mediaries between these two sides can untie the knot if they correctly combine the transport and logistics solution with the Lo interval, that is, from the warehouse of the supplier to the FOB or CIF financing. This is a task containing great risk, since both the buyer and the seller may appear as an untrustworthy partner, but a rational decision may be thwarted by some unexpected obstacle, i.e. the cancellation of export license with the change of government or its members of the exporting country. In a similar way, many of our countries' businesses have suddenly collapsed after only a few successful years.

The conclusion which can be drawn from the last points is that the most important prerequisite for successful co-operation is bilateral trust and the accumulation of experience, civilised development of direct ties, their continuity and protection from haphazard, criminal and corruptive forces.

The flowing waters must clean themselves. Let us hope that the dynamics of the second market and the new economic activity in our countries will help rid all of these evils and develop normal market relations and co-operative ties.

THE PROBLEMS AND PROSPECTS FOR LITHUANIAN SCIENTIFIC-TECHNICAL CO-OPERATION IN DEVELOPING INDUSTRY AND INFRASTRUCTURE

ADOLFAS ŠLEŽEVIČIUS

On behalf of the Government of Lithuania, I would first like to congratulate our honoured President K.Prunskienė and all ladies and gentlemen who have gathered here in Vilnius, at this very significant conference. Before starting on the topic of this conference, I would like to say a few words about the general political and economic situation in Lithuania, without which, understandably, it would be difficult to comprehend the currently events in our country in the mentioned area and our tasks for the nearest future.

I would first of all like to draw the attention of all our guests and scientists here today to the fact that we find ourselves on the eve of a very important event for Lithuania. And the topic, which was analysed at this conference, is directly linked to this event, that is, the foreseen signing of the European Agreement on June 12, between the Baltic States and the European Union. In essence, this is a very significant event in all aspects in the history of independent Lithuania, since our country, like our Baltic sisters, is moving from one political space to a completely new one. Of course, needs and benefits are not the only factors which await us upon the receipt of EU Associate Member status. Larger tasks await us as well, and in general, we face much work and many problems, for which we are actively preparing. In the Ministry of Foreign Affairs we are in the process of establishing a special Department for European Integration, with special divisions in each ministry, which currently should concentrate their efforts in examining the so-called EU White Book, which will define all priorities and the road Lithuania must take during the transitional period to full membership.

We possess ambitious plans, which in my view are realistic, for which we will all work together in agreement to become a full-status member of the EU by the end of this millennium. This depends on the development of our political and economic reforms, the methods we employ in fulfilling those tasks, in normalising or correlating the legal and normative acts which are currently in effect in our country, with the accepted legal basis of the EU. This aspect, I feel you will agree, is related to another important aspect - our co-operation with the EU defence or security structures in the military sector. As you may well know, Lithuanian President Brazauskas expressed the principal statute of our country at the beginning of last year - our striving toward full and equal NATO membership. In

K. Prunskienė and E. Altvater (eds.), Transformation, Co-operation, and Conversion, 13-18.

comprehending our national geopolitical situation, the reason for our desire becomes very clear and easily explained, since we are a small state, and because we cannot guarantee our security through our own forces. Thus a collective conception of security is very acceptable to us and we are very pleased that a large part of Western Europe and the US support us and our basic position with understanding. We were very pleased to learn that during the meeting of many high-ranking government officials from the EU and visiting US Vice-President Al Gore in Tallinn, it was constituted that NATO expansion is a process, that there are no lists of countries, spelling out priorities, there are no drawn boundaries, since it is understood that there would be nothing worse for us than an undefined situation.

We understand as well that to attain this goal we must also expend all efforts in all spheres of activity to unify our standards, capacities, methods, etc. Thus we very actively participate in and were among the first to sign the Partnership in the Name of Peace Program, after which intensive work followed. We prepared an individual "partnership in the name of peace" implementation programme which we co-ordinated, adopted and financed. Of course, the ongoing implementation of this programme is a very significant step toward bringing us closer to European security structures. Along with our Baltic neighbours we have also established and actively participated in other programmes.

The EU process of economic and defence integration is similar to our integration process with our neighbouring countries. We are quite pleased that during the last two years we have cemented our co-operation on a pragmatic government level with our Baltic neighbours, first of all Latvia and Estonia. As you most likely know, since September 13, 1994, we have a rather functional Baltic Council of Ministers, which we established based on the experience of the Scandinavian Council of Ministers. Currently 19 committees operate within the framework of the Baltic Council of Ministers; these committees work out issues which are directly related to the topics of this conference - the co-ordination of various areas, beginning with air traffic control and border control, transport, customs co-operation, energetics and other spheres. Although the work of these committees may not be fully effective yet, I feel that these main directives are defined and may be used as a tool for co-ordinating such co-operation on a government level. The Baltic Assembly established on the parliamentary level is functioning rather efficiently. It is very important that the activity of the Baltic Council of Ministers is progressively co-ordinated with the Northern Council of Ministers, and that the co-operation of the eight Baltic countries is applied to solving specific co-operation issues, for example, the successful decision making on economic co-operation in the region and the question of security guarantees. Interest here is quite widespread and will undoubtedly lead to many future discussions.

In this context it is also pertinent that we maintain favourable relations with our neighbours, which we steadily seek. We have signed a bilateral political agreement with neighbouring Poland, which opens the doors to further economic co-operation. With Belarus we signed a border agreement. Bearing in mind that we have never had a border between our states, this is an especially rewarding accomplishment. Let us not forget Russia as well, with which our efforts resulted in favourable normal relations. This is a very specific and complex issue, remembering the situation of the Kaliningrad region with a high military concentration. This issue also deserves an especially subtle discussion, and we are pleased that we are in the process of establishing methods with the Russian government for settling the military transit issue, which is currently taking place according to the rules and regulations agreed upon by both nations. In its turn, Russia ratified a favourable trade regime agreement, which is also notable, since, although the general volume of trade with Russia is decreasing, it still remains our largest trading partner, comprising about 30% of our total trade volume. We also understand Russian concerns and its constant affirmation that NATO expansion poses certain problems for Europe from the security aspect, and upon evaluating these concerns, we in our turn stress our principle convictions regarding our long term goal to become a member of this organisation as well. We also seek that the process of European security and its stability is not violated; and it must be stressed that this is also a long term process, requiring much work from our side for its realisation.

In speaking about the possibilities for Lithuanian scientific and technical co-operation, the restructuring of our science and industry and military industry, a few words should be said about our national scientific potential, which, up until Lithuanian independence, was rather concentrated and mighty. These former scientific research organisations exist today as well, however their financial situation, for understandable reasons, has worsened, due to the large economic decline during our first years of independence. I would like to provide some data on our research potential.

Our country has 29 national scientific institutes, which employ 200 habilitated doctors, 1400 doctors and 2200 scientific researchers. We have 15 national institutes of higher education, in which 450 habilitated doctors, 3500 doctors and 5500 scientific pedagogical instructors conduct research and teach students. We also have 15 national scientific establishments in which 50 habilitated doctors, 260 doctors and 500 scientific workers conduct their work. According to these numbers, our scientific potential is rather high. As I mentioned, total financing of these institutions has decreased, however, in considering the importance of science and education, in our economic and societal development, our government constantly increases the percent of allocated funds from the national budget. In 1993, 5.7% of the national budget was allocated toward these sectors, in 1994, this was increased to 6.1%, and in 1995, 7% of all funds was allocated to these spheres.

It is apprehensible, and I must stress that, although tendencies are positive and the percent of budget funds have been increased, this financing is still not sufficient for these institutions. For this reason we seek other possibilities for encouraging financing of various programmes of study. In 1993, we established a national scientific studies fund for the financing of the best, most prospective and fundamental scientific research. Along with the UNDP and the Lithuanian Stock

Innovation Bank, we are establishing a Lithuanian Innovations Centre with a separate fund for the preparation of scientific research. Support for participation by scientific studies institutes and separate scientific groups in international science and study programmes is granted by such programmes as Tempus, Eureka, Peko and others. Along with the Soros foundation, we established a special fund for the encouragement of individual scientists as well.

During the last few years Lithuanian scientists participated in various international research projects: 11 Peko EU research and studies projects, 15 European Commission Copernicus projects, 8 economic scientific co-operation ASI projects and over twenty projects within the TEMPUS/PHARE framework. This year we have also increased the budget for PHARE TEMPUS financing almost twofold - it now totals 3.5 mln ECU.

Currently, approximately 10 percent of all PHARE funds are allocated to the TEMPUS programme. Our task now remains to find all possible means to support and finance sciences, the organisation of various programmes and their preparation and implementation. We also comprehend that a very important precondition lies before us, and this is the more advanced education of our younger generations through the education system. The Lithuanian Government has accepted the decision, and the Ministry of higher education has announced a contest, to organise computer classes in all Lithuanian schools. This will add to the educational preparation of our youth.

The participation of scientists in government programmes plays a vital role. The government has accepted an array of large complex programmes, including very important state economic spheres, such as the transport programme, energy, agriculture and other programmes, in the preparation of which scientists from our country and abroad participated. We are grateful for the support which we received from the EU in various forms in preparing our specialists and in consulting, in the preparation of various programmes, for example, the financing and realisation of the national transport programme, which necessitated about 3.5 bln. USD. This programme joins the European standard railway from the Polish border to Kaunas and further north, upon restructuring and significantly increasing the capacity of the Klaipėda Port almost double, preparing many specialised terminals, which are now non-existent, such as ro-ro passenger containers and liquid gas and other goods terminals. We have also, in co-operation with our scientists, completed research and prepared the Via Baltica plan, which will increasingly play a significant role in our national economy, the so-called North-South corridor. This project alone will require financing of a few hundred million dollars, for its full development within the infrastructure. Such projects are not possible without consistent co-operation with regional scientific research organisations. Many scientists have actively worked in preparing these programmes, such as energy, which has special importance in our country; an area requiring the increased attention of scientists is the safety check of the Ignalina AEP - our scientists along with international experts are doing much in this field to guarantee the safety of this especially dangerous object. Funds received in the form of credits also help in this area, as well as the grant from the EU through the European Bank of Reconstruction and Development in the sum of 40 mln USD, to increase the safety of the atomic plant within the framework of the energy programme. An array of other important topics have been analysed and formulated, such as the increase of hydroelectric capacity, the expansion of an electric bridge to Poland, to enable integration into the European energy system , as well as energy-saving technologies - here, the help of scientists is needed in heating and other various energy object reconstruction. A very important topic, which is very real to us, is the reform of the agricultural system and the help of scientists in the preparation of the national agriculture programme; we have many sub-programmes which analyse agricultural and reform issues, questions of economic mechanisms, market security, new technology transfer and many other topics. We may thus say that truly this topic on the development of scientific-technical co-operation among our countries is a great, important topic, on a national level and for commercial industry as well.

In speaking of conversion, a few words may also be said on the issue of conversion of the enterprises of the former Union military complex in our nation. We used to have many working objects in the electronics industry. The current situation is not identical for all of them, however, it should be said that through the rapid privatisation conducted by our government, these enterprises are finding their niche in various markets, the search for which is important not only in the area of individual military complexes but in the economy in general. It should be noted that about 63% of the gross domestic product is comprised of the private sector. At the end of the year, we expect this to reach 70% of the GDP, quite an important element in the economic reform process. I would also like to note that last year, for the first time in the five years of our independence, GDP began to grow. According to data from the Department of Statistics, the growth rate is 6%. IMF and World Bank statistics are more optimistic, however it could be said that on the basis of privatisation and other economic reforms, the worst times of our economy are in the past. This year we expect a 5% increase in GDP, a prognosis which we presented to the Seimas, upon the confirmation of next year's budget. Experts of our financial institutions plan a somewhat higher increase - 7%. We understand that a higher increase will only be beneficial to our economy. Thus, conversion and the general restructurisation of industry increasingly gain an irrevocable tendency, and this is a guarantee for our general reform and economic growth. What is especially important is that Lithuania, as all other post-communist states with certain preconditions, during those 5 years, truly survived large social and economic crises, and there is nothing more important than the guarantee of a rapid economic growth rate, and on that basis a solution to the main social problems Lithuania is facing. This is a very important task for our government, which constantly pays attention to it, along with the potential of our scientific institutions, which also work for this goal, including the institute headed by K.Prunskienė, which truly does much to ensure that all of these ideas are presented to our government and implemented in our state. These are some main ideas, in brief, on this very important issue.

LITHUANIAN SCIENTIFIC POTENTIAL: NEW TASKS AND PROBLEMS OF EUROPEAN INTEGRATION

V. DOMARKAS

Upon redeeming its independence, Lithuania inherited a comparatively well-developed science sector. It is enough to note that currently there are about ten thousand scientific doctors in Lithuania, a majority of whom still conduct scientific research and pedagogical work in fifteen institutes of higher learning, twenty nine public scientific institutes and fifteen public scientific establishments. The level of Lithuanian science is not uniform throughout all areas and as a whole does not meet European standards, however if we single out the qualifications of scientific doctors, it should be said that recently the level of Lithuanian scientific groups and even scientific institutes is equivalent to that of Europe.

Cardinal changes in the political and economic system of Lithuania created completely new functioning conditions for Lithuanian science. It is no secret that just recently a majority of scientific research and even the continuous work of scientific institutes were linked to the central structures of the Soviet Union, including military-industrial complexes. With the disappearance of orders for those types of complexes, funds from the Lithuanian budget alone do not cover the previous levels of financing for science, and this is unfavourably reflected in the state of fundamental and applied research. Secondly, the decline of the scientific sphere is not correlated with the establishment of an economically developed nation, based on democratic principles. Thus, in this stage it is important to secure Lithuanian scientific potential and its effective utilisation. The significance in securing this potential lies as much in internal scientific policy as well as in international support.

1. The Status of Lithuanian Science

Internal political decisions are dependent on the understanding of existing economic reality and scientific functions. Bearing in mind that Lithuania is not wealthy in raw materials, its only valuable resource may become the intellectual and professional level of its inhabitants, which would guarantee international achievement in science as well as the participation in modern technology-supported production. It is thus expedient that scientific research is supported by government subsidies and individual economic investments. This year almost 7% of the

K. Prunskienė and E. Altvater (eds.), Transformation, Co-operation, and Conversion, 19-23.

government budget expenditures were allocated to science and studies, of which about 17% is used in financing scientific institutions. These funds satisfy only the most minimal demands of scientific institutes.

Under market economy conditions, technological advancement depends on scientific potential as well as competition among economic subjects. Thus economic subjects should also be interested in the development of applied science. The current financial situation of Lithuanian state industrial enterprises, however, is not in good shape, and these cannot direct adequate attention to the creation of new technologies or the perfection of obsolete technologies as well as supporting Lithuanian applied science. The expanding private sector in industry, in my opinion, has only begun to grasp the importance of technological innovations, thus in the future we may only expect significant support for Lithuanian science from private capital structures. It is for this reason that the problem of rationally distributing government funding for scientific research bears much importance in this stage.

Until recently, scientific institutes were financed according to the number of their employees and not their scientific productivity. Thus in furthering Lithuanian scientific reform, it is mandatory to take into account the urgency to the level and state of the scientific research conducted in each scientific area, in each scientific research and study centre. The first stage of an evaluation of scientific and study institutions conducted last year, was unfortunately not sufficient for an objective assessment of all scientific and study institutions. This year the Norwegian Scientific Council of experts is conducting the second stage of their evaluation. We hope that the results of this study will provide an ability of better distribution of financing among scientific institutes. This international evaluation of scientific and study institutions also has significance for the international integration of Lithuanian science.

2. International Support and Joint Projects

As I have already mentioned, the most important task during this transitional period is the preservation of scientific potential. International support through providing Lithuanian research with national (i.e. the European Union) or private (i.e. the International Science Fund) financing has been significant. Such financing is important for two reasons. First, these were the first cases in which our scientists participated in competitions and in which an international examination was conducted of submitted projects, all of which was meaningful in reorienting to world-wide accepted criteria for evaluation and aided in distinguishing top-notch scientists. The best scientists also received financial support for scientific research, which in this case secured them in the scientific sphere. This year scientists who received assistance from the International Scientific Fund are also supported by a co-founded 2 mln. Litas foundation of the International Scientific Fund and the Government of the Republic of Lithuania.

The mentioned first attempts by Lithuanian scientists at competitions were not for granted. The scientists received experience as well as confidence in their abilities and are continually more actively and, what is especially pleasing, successfully trying to participate in various EU sponsored projects with EU scientists. Lithuanian scientists have participated and are participating in:

- 11 PEKO (the third EU scientific and research programme) projects;
- 15 EC Commission COPERNICUS programme projects;
- 8 projects in the Action for Co-operation in Economics - ACE programme;
- a few projects in the NATO scientific research programmes;
- this year two new EUREKA projects are being realised;
- TEMPUS projects assisted scientists in higher education institutions.

For Lithuanian scientist groups participating in common projects with foreign scientists, it would be beneficial to grant government financing priorities, since just the fact of being accepted in these projects is a positive international rating for the quality and urgency of the activity of Lithuanian scientific teams. On the other hand, it would be beneficial to request the EU and the Northern Council of Ministers, etc. to analyse the possibilities of establishing a fund for scientific research support in Lithuania, allocating specific funds for a period of three to five years and dividing them according to international project evaluation. In my view, a yearly grant of 1-2 mil. USD would stabilise the situation of Lithuanian science as well as create new stimuli for attracting youth to science. In striving for such support, it should be mentioned that this is directed toward the preservation of scientific potential. All scientists at the international level are important to us, and there are not many of them.

3. Other Funding

Other possibilities exist for realising international support through combining the funding by other sources. Mr. G. Soros has stated that he is prepared to donate a percent to such joint funds for fundamental research in support of the natural sciences, which would be created according to the following:

- 50% from the country seeking support,
- 25% G. Soros,
- 25% private or state sources from Western countries.

It should be specified whether the EU or Northern countries could participate in the creation of such a fund. Undoubtedly the first thing needed is the decision by the government of the Lithuanian Republic regarding the division and size of such funds.

4. Local Development and Support

On the other hand, it must be said that there are also other tasks which, in administering Lithuanian scientific reform, could be accomplished by Lithuanian academia without the help of the international community. At the top of the list are the encouragement of scientific and studies integration, from a practical sense, not just from an administrative perspective. Scientists from state scientific institutes must more actively participate in the studies process, through which the professorate of higher education establishments would be given more time to spend on research. Without doubt this would increase the teaching level in higher education institutes. The academic community must realise the benefit and pertinence of this process.

Another problem is the reorientation of science. A large range of scientific fields exists, in which, disregarding their strategic meaning, there are no or almost no specialists in Lithuania prepared to conduct scientific research on an international level. If we do not conduct such work, we will lag behind world progress in comparable fields, which will result in lowered specialist preparation in Lithuania. Lithuanian scientists should more actively reorient and conduct scientific research pertaining to Lithuania. All of this would require a lot of funding, however, taking into account the wide range of Lithuanian scientific interests, in separate instances such a reorientation could be painless and would not require much additional investment. In my view, there would be many such cases. It is also necessary to definitely decide on priorities in the "traditional" Lithuanian scientific spheres as well. In applied science this is necessary for joining EU projects.

In assessing given scientific potential and the importance to Lithuania of developing equivalent branches of science, the Expert Committee of the Ministry of Education and Science recommended that the following areas of applied science be given priority:

1. Ecology
2. Biomedicine
3. Biotechnology
4. Information and Communication Technology
5. Energy
6. Agricultural Science and Food Technology
7. Metrology, Measurement and Quality Assessment.

Members of the aforementioned Committee include: the President of the Rector Conference, the President of the Lithuanian Academy of Sciences, the Head of the Scientific Council and other distinguished Lithuanian scientists.

Lithuania has almost no specialists qualified to consult patenting issues and analyse existing patents. Consultative support from the West is still too expensive. Without the possibility of receiving international level patent consulting, we potentially lose the possibility for patents; scientists work without knowing whether their work is being repeated and thus cannot objectively evaluate to what extent their work does not violate existing patents, etc. International assistance here is mandatory, both short-run assistance by hiring foreign patent consultants, and long-run assistance in preparing Lithuanian specialists.

Analogous problems and demands exist in preparing scientific marketing specialists capable of administering assistance in market analysis, evaluating certain innovations based on planned or completed research, market perspectives, possible prices, etc. It is expected that significant aid for Lithuanian scientists in these areas will be granted by the Ministry of Education and Science, programmes of the UN as well as the Lithuanian Innovations Centre which is being established by the Lithuanian Stock Innovation Bank.

A method in arousing the interest of foreign business people or scientific research leaders (in national or private institutions) is the use of the cheap labour of our qualified scientists, constructors and manufacturers. Our citizens are incapable of seeking such contacts without knowledge of world market demands for such labour. Foreign partners, who know our situation and have evaluated our qualifications, could certainly establish tasks and control their fulfilment, including orders placed for scientific research. The benefits would be bilateral, as Lithuania would at the same time be preserving its scientific potential. Such co-operation could incorporate both small contractual assignments and long-term contractual research.

5. Concluding Remarks

In conclusion, having compared the high potential of Lithuanian science, I would like to again stress that it is necessary not only to preserve it but also to effectively utilise it, not simply for the sake of Lithuania but for all of Europe. The Lithuanian academic community must comprehend and, in my opinion, does comprehend the importance of Lithuanian scientific reform and reorientation. Lithuanian science must find its place among the European scientific system. The speed at which this occurs depends not as much on our government, the Ministry of Education and Science or international organisations as on scientific and study centres, scientific teams and each and every scientist.

CITY PLANNING IN A NEW GEOPOLITICAL SITUATION: THE CASE OF VILNIUS

DALIA BARDAUSKIENĖ

Co-operation of different states in the know-how sphere is a topical issue today. Changes in the geopolitical and economic situation of different countries and regions call for new ways of acting in all spheres of life. Therefore, this is a central issue in Spatial/Regional Planning as well.

So far, Lithuania does not have its own national programme for regional spatial planning and is in the process of creating a legal base for Territorial Planning, as well as the replanning of settlements and towns.

Much work is done in an international context. We are still learning the planning methods of a market economy under democratic conditions. The principles of public participation and subsidiarity are just being introduced now. The municipality, the public and business people are playing new unfamiliar roles, roles which provide us with experience and understanding, encouraging the planning of Lithuanian territory in such a way as to attract investment, ensure community stability, and maintain the sustainability and uniqueness of the environment.

I would like to discuss the creation of the "Vilnius General Plan: Vision Two Thousand and Fifteen (2015)" and the ways in which we have gained from international experience. I will specify our procedures in practice and the results achieved.

The last Soviet period City Master Plan (called the "Genplan") was completed and approved in 1978. It provided for city development until the year 2000. The city was supposed to have about 700 000 people. A suburban zone was planned around the city which:

- provided the city with agricultural products;
- provided sites for the development of collective gardens and small resting houses;
- served as a rest and recreation area for inhabitants;
- provided for the disposal of city garbage and other waste materials;
- provided a source of work power.

Planning was a very simple issue in those times. Territorial city development was given priority. Planning itself was passive, the process was very clear and simple - as it was directed from above, it was centralised. Its primary goal

K. Prunskienė and E. Altvater (eds.), Transformation, Co-operation, and Conversion, 25-30.

was the satisfaction of production needs. The following criteria were looked upon as secondary:

- size of the labour force
- size of the population
- need for living space
- need for services.

City planning and plan making were basically looked after by the central governmental institutions that decided on:

- which cities needed planning,
- individuals responsible for drafting designs,
- distribution of funds.

Municipalities used to take part in this planning process as well, as they had to approve the final product, i.e. Genplan. Nevertheless, the actions of municipal bodies were only formal procedures.

After regaining independence in 1990, the Lithuanian Parliament passed a Law of Local Self-government which stated that municipalities should prepare and approve general schemes for city and town development, as well as organise their implementation.

Pursuant to law, the Vilnius Municipality undertook the territorial development of the city. The neighbouring municipalities also operated according to the same Law of Self-government, but very often they ignored the suggestions of the city of Vilnius. Only by the usual means, that is "by directing from above", the 2,500 sites were given to the Vilnius City inhabitants for the development of single apartment buildings in the suburbs of the city. There was no room for the development of such apartment-buildings on city territory.

All attempts to legitimise new city boundaries were failures. The same experience of larger Lithuanian cities showed that physical planning on the regional level acquired negative connotations. This process placed city and country at odds with each other. Moreover, territorial plans and planning processes have lost their prestige within the municipalities.

The reason for this is that the new territorial planning system is still under development in Lithuania. The Land Usage Reform and restoration of real estate is being implemented in the country, while the legal base for territorial planning has not yet been established.

The debate over the meaning of democracy in the physical planning process has not been settled. A planning philosophy that incorporates the principles of transparency, reliability and participation is still in the stage of discussion. The majority thinks that democracy equals total freedom of action with no rules or discipline. Unfortunately, we have quite a few examples of such "democracy".

1. The Present Situation

Under such unfavourable conditions, the Vilnius Municipality initiated the development of a new City General Plan.

Why does Vilnius need a new General Plan? There have been substantial changes in the political and economic environment. Vilnius has become the capital of a European country. There is a need:

1. to assess Soviet inheritance and create a new conception for city development under the conditions of competition;
2. to create the image of a European state capital;
3. to assess the consequences of urban sprawl;
4. to solve engineering, infrastructure and transportation problems created by "sprawled" territories;
5. to liquidate discrepancies between the bedroom communities, work places and the distribution of services;
6. finally, to set up priorities for the territorial development process by 2005.

In 1991, a task for the "General Plan of Vilnius - 2015" was approved. The designing institute "Urbanistika," being the only institution of its kind in Vilnius, was assigned to carry out all design work. Conditions and restrictions on the design work have been issued by different ministries and departments in charge of the planning processes.

Following the specially prepared programme for a background study, a number of very different and large-scope research studies have been performed:

- an assessment of the geological and ecological situation of Vilnius and its suburbs;
- a cultural heritage study;
- a study of the transport situation;
- an assessment of the engineering infrastructure;
- a bioenergetic study of the territory;
- an assessment of industrial territories and public utilities;
- a social study, as well as other studies.

A total of 17 different research projects were conducted.

After the completion of all these research studies, a competition for the preparation of the conception for territorial city development was announced. The outstanding urban planners of the Vilnius Technical University, as well as the Vilnius Art Academy and the designing firm "Urbanistika" took part in the competition. As a result of the competition, 4 (four) variants of the conception for Vilnius city development were produced:

- Concept of Dispersal (which is topical in the city now)
- Concept of Concentration (all new growth should move to the south-east of the city)

- Directional Conception (development of the Vilnius -Kaunas corridor)

All material of the competition was assessed by the Council of Independent Experts. The experts drew conclusions and selected the best proposals. On the basis of competition material and papers prepared by experts, the Municipality of Vilnius prepared a very important document: "Strategy For The Spatial Development of the City of Vilnius: Program for the Concept of the City Official Plan".

Despite the immense work thus far performed, these documents did not answer the following question: What is the mechanism for the economic development of the city? It contained no clearly stated directions and goals for the city to guarantee successful city development.

In 1994, while the programme for territorial city development was being formulated, the Canadian Government provided the City of Vilnius with financial assistance through the Canada-Baltic municipal assistance programme "Strategic Planning for Municipal and Environmental Management". This programme focuses on the strategic planning of cities under the conditions of democracy. The city government agreed that Vilnius needs a comprehensive and integrated strategic plan based on public participation and input.

Why does Vilnius need strategic planning? The city must:

- define effective directions for city development;
- ensure the rational use of limited resources;
- balance interests of different social groups;
- encourage investment;
- implement self-government and democracy within the city;
- set conditions for community involvement in city management and future planning.

In December 1994, the Conference of Ministers from the Baltic Region prepared a document entitled "Vision and Strategies around the Baltic Sea 2010." This important document spelled out the direction of city development in the Baltic Sea region and called for local initiative and response. Cities in this document are divided into four functional categories:

2. The Functional Classification of Cities

European cities with high-ranking international functions
National cities with important national functions
Regional urban centres
Other major cities

In this classification, the city of Vilnius is placed in the third category as a regional urban centre. However, in the vision part of this document, Vilnius is placed in the second category as the "National city" with important national functions. Following this document and preparing the Vision document for Vilnius 2015, we must prepare the city to attract:

- embassies and consulates
- major international organisations
- international fairs and exhibitions
- international conferences and seminars
- international tourism

Under the new directions from BSR Vision 2010 and the Canadian Strategic Planning Program, the organisational structure for this process has been established. The SWOT (strength, weaknesses, opportunities and threats) analysis of the City of Vilnius was conducted, providing for 3 (three) possible options for city development:
The option of Vilnius as a centre of culture, science, and tourism.
The option of Vilnius as a centre of business, communication, and commerce.
The option of Vilnius as a centre of commerce and industry.

The conceptual part of the City Official Plan (basic drawings and programmes) has been discussed with the public following the rules of order No 104, 1994 05 04, issued by the Ministry of Construction and Urban Development, the Lithuanian Republic, and the decree of the Vilnius Mayor.

This was the first time a trial of such scale was ever carried out in Lithuania, involving all possible means of informing the public and involving it in the planning process. The Vilnius Office of the CUI has prepared a special programme to serve this process. The programme of Public Consultation of the conceptual part of the Vilnius Official Plan entailed the following means and tools:

1. informing the public through mass media;
2. an exhibition of the Vilnius Official Plan from March 14 to April 4 in the Museum of Applied Arts;
3. an International Workshop "Strategy for the City Development" for local and foreign experts;
4. two major meetings of the Official Plan makers with the citizens of Vilnius;
5. public opinions that have been registered in the journal during the exhibition in writing; orally - the citizens have been expressing their opinions by telephone which were continually announced in the mass media and placed in the Vilnius Municipality (62 50 55).

The total number of citizens providing their opinions and acquainting themselves with the Official Plan is about 2000, 150 of whom were experts. The Ministries of Lithuania presented their opinions following their rule and order, that is, in "conclusions or statement of co-ordination", as they call it. The following conclusions have been drawn from this process:

Discussion of the Vilnius Official Plan conception with the public showed that the attitude of the urban planners to the public has changed. Nevertheless, Soviet traditional thinking is still alive (when official plans were secret and accessible only to the "selected" circle, and their realisation had been handled by the Government and its centralised budgeting). Now planners view public hearing of the planning documents as giving the information for free, and as interference in their personal monopoly of "authorised work".

Public participation is a necessary alternative measure against former bureaucratic centralised planning.

Public participation in planning is not equivalent to an input, especially in the very beginning, when all the process members have no experience in the process they started. Not all of the participants have enough knowledge and information about territorial planning.

Work with the public means constant work with the representatives of mass media, without allowing them to monopolise the sphere.

By involving the public in the planning process, we may alter the current image of plans and planning activities from the "forbidden" and "restrictive" to "helpful", "encouraging", "effective".

At present, the conception of The Official Vilnius City Plan and its variants are in the stage of being completed. The new City Council (elected by the public) will choose the City Vision, and this will be a guideline for the territorial development of Vilnius. Their choice will reflect local needs and opportunities that are prerequisites for the generation of working places, the need for a living space and the provision of services.

I would like to express my hope that The Official Vilnius City Plan - Vision 2015 is the best way of utilising and regulating city resources. The new Land-Usage Plan will encourage investments and provide for reasonable freedom of action.

3. Conclusions

International assistance in the area of territorial planning is very important and useful, enabling our professionals and specialists to exchange and receive information and knowledge on planning under the conditions of market economy and democracy. So far, availability of Western know-how in this area has been sadly lacking. Where available, it has been limited to ministries and large urban centres. But such know-how is absolutely necessary for all of the 56 municipalities and 10 counties, if progress in planning is to be achieved and Lithuanian territorial planning is to meet Western standards.

CONVERSION OF MILITARY INDUSTRY AND OTHER MILITARY FACILITIES IN LATVIA

JURIS PRIKULIS

During and after the collapse of the Soviet regime in Latvia, the military industry belonged to those parts of the established system which were most affected by these events.

Riga was the centre of the Baltic military district (which included Latvia, Lithuania, Estonia and the Kaliningrad region of the Russian Federation) of the Soviet armed forces, and this considerably influenced the increase in the volume of military production in Latvia during the entire period of Soviet rule. It is believed that the share of industrial output of military items was larger in Latvia than in neighbouring Estonia and Lithuania. In 1995, the total volume of manufacturing (including civilian) in Latvia, in comparison with 1990, constituted only about one third. A substantial part of such decline must be attributed to the almost complete interruption of military production. Soviet (Russian) army withdrawal contributed to military production decline as well.

After the occupation of Latvia and its incorporation into the USSR in 1940, the Soviet leadership rapidly increased its military presence in Latvia. After World War II (WW2), this process continued. Between 1940 and 1994 (when Russia withdrew its armed forces), more than 3 thousand Soviet (Russian) military units were stationed in Latvia. In many cases, in the process of establishing military sites and facilities, the interests of the local population were violated in very brutal forms, and heavy material losses were inflicted on local enterprises and individuals.

As Latvia almost directly bordered the West, the Soviet leadership in Latvia, instead of manufacturing guns and tanks and the like, chose to produce other necessary equipment for the army and components of various military machines, apparatuses and devices. Especially important in the eyes of the Soviet military were Latvian factories, which produced radios, electric machinery, telecommunications equipment, motors and instruments. As a rule, those factories were subordinate to the all-Union ministries, and the local Latvian SSR government had little say in these matters.

Beginning with the occupation of Latvia in 1940, civilian production factories were forced to open "secret shops" for the production of military goods for the Soviet Armed forces. At present it is difficult to have a full picture of exactly how many factories in Latvia possessed such equipment for military production.

K. Prunskienė and E. Altvater (eds.), Transformation, Co-operation, and Conversion, 31-34.

Military production was conducted, for example, in agricultural machinery factories and experimental scientific research facilities.

It is widely accepted that three large electronics firms, namely VEF, Komutators and Alfa, were heavily involved in the production of goods for the Soviet military forces. As an example, let us take a look at the development of old and new factories in this sphere. VEF is one of the oldest and largest factories, its origins dating to before World War I (WW1). Before the Soviet occupation, the radios, telephones and similar equipment produced by this factory were quite well-known. After WW2, it became one of the major suppliers of such production for the entire USSR, exporting its civilian production to the Soviet satellite countries and third world countries. VEF became the leading producer of telecommunications systems for the COMECON countries. In the late 1980s, about 22 thousand persons worked in this factory. A substantial part of VEF's production was manufactured in accordance with direct and, more often, indirect (for example, through the Ministry of Communications of the USSR) orders from the Soviet Ministry of Defence.

In 1961, part of the VEF military branch served as the basis for a new factory, which, in contrast to the VEF, was predominantly a military factory called Komutators. In 1971, on the basis of the Riga Semiconductor Factory, the new military-oriented electronics factory Alfa was established and became one of the largest factories of such kind in Europe. These two factories also produced some civilian goods but these did not comprise a very essential volume of their total output.

Another group of military factories was directly subordinate to the Soviet military forces. Part of the military production of these factories on the territory of Latvia was connected with maintenance (tank repair, ships and other military equipment). Some of these factories produced various auxiliary items for the army (metal forms for construction, concrete panels, foundation blocks, road asphalt, etc.).

It is estimated that in Latvia more than 100 thousand individuals were involved in military production and research in the early 1980s in establishments which were not directly subordinate to the Soviet army. A majority of these persons were post-war newcomers from Russia and other former republics of the USSR.

The conversion of military factories and military facilities in Latvia took place in several stages. During the Gorbachev era, some modest conversion plans were initiated, but remained, in general, unfulfilled. The Soviet military leadership began withdrawing the production of some military goods from such rebellious republics as Latvia, and as a result, military production in Latvia began to decline substantially.

In 1990 and 1991, during the final stages of Latvian re-establishment of independence, military production experienced a dramatic decline. After the restoration of independence, this decline in military production orders for factories in Latvia continued.

The Riga diesel engine factory, for example, was gradually losing orders from the Russian military. If these orders comprised more than half of production in 1993, then there will be almost nothing left in 1996. The factory is partially successfully seeking military orders as a replacement.

During the Gorbachev perestroika, the establishment of co-operatives connected with existing state-owned factories was allowed, including military factories. Co-operatives associated with military plants began producing civilian goods and services, in this way helping to convert some capacities used for military purposes to civilian ones. A number of employees of these factories left and became members of co-operatives.

In many cases, part of the equipment and other resources (raw materials, etc.) of military production factories was sold to co-operatives. This was often done in violation of Soviet laws. When the authorities of the Republic of Latvia increased their command of the situation, they issued regulations regarding the handling of problems connected with conversion of military factories and property.

A draft programme for the conversion of military facilities to civilian use was presented for parliamentary debate in April 1991. The work of the military conversion commission, established by the Council of Ministers on December 17, 1991, was countered by strong resistance from the ex-Soviet military. There were some conflicts between the Latvian and Russian authorities in the initial phases, especially over procedural matters.

In September 1992, the Latvian Parliament passed a law which placed former Soviet military facilities under the control of local governments.

The rapidly changing environment did not allow for elaboration on a sufficiently realistic programme for the conversion of former military facilities.

There were many unrealistic expectations about the value and possible use of the former military facilities. In 1992, ministries in Latvia submitted proposals to grant them the right to use these facilities but soon abandoned many of their applications. In many cases these facilities became a heavy burden to local governments. Part of the former military facilities were plundered by Russian army personnel or later by local inhabitants. The huge problems in converting former military facilities are connected with the ecological damage inflicted by Soviet army units. At the same time, many examples exist elsewhere in Latvia in which the local governments and private entrepreneurs are successfully converting and utilising the former military facilities.

To a larger extent, factories which were basically producing military or military-connected products, are in a very difficult situation. For example, in Alfa, the number of employees has decreased about ten times and production volumes have decreased to an even greater extent. In Komutators, this decrease has not been as steep. These and other military related factories are seeking new purchasers, investments, and in certain aspects they have achieved some success.

One of the most urgent problems are the retraining of former Russian-speaking employees of military factories and granting financial assistance to those of them who have decided to leave for Russia or other CIS countries but are not able to do so, as they lack the necessary financial resources.

Among the obstacles especially felt in the military factories on their way to restructurisation is their lack of experience in marketing the new products which are being developed for sale. They are often unable to prepare adequate business plans and establish fruitful contacts with foreign investors.

Former military factories are currently under the process of privatisation - some are privatised already.

One of the peculiarities of the process of privatisation of the former military enterprises, which were under the direct control of the Soviet armed forces, is that in accordance with an agreement between Latvia and Russia, real estate property in these factories was declared Latvian, while other capital (equipment) was declared Russian property. The Russian government provided the possibility for the employers of their factories to rent this capital and after certain procedures obtain ownership. It seems that Latvian authorities have found proper ways of proceeding in these affairs.

It seems clear that in the foreseeable future of Latvia military-related production will be significantly reduced - Latvia has already shown its interest in other directions of industrial development.

References

1. Kristapsons, J. Tjunina, E. Kalinina, G. (1995) Transformation des Wissenschaftssystems Lettlands (1989-1993), in R. Mayntz, U. Schimank and P. Weingart (eds), Transformation mittel- und osteuropäischer Wissenschaftssysteme. Länderberichte, Leske-Budrich, Opladen, pp. 88-174.
2. Latvijas statistikas gadagramate. Statistical yearbook of Latvia (1995), Latvijas Republikas Valsts statistikas kometeja, Riga, p. 37.
3. Rjekstins, J. (1995) Izpostita zeme. PSRS okupacijas armijas nodaritie zaudejumi Latvijas kulturvide, Latvijas Valsts archivs, Riga.
4. Upmalis, I., Pavilionis L., Berzins, V. (1994) Vini aiziet, objekti paliek, in Neatkariga Rita Avize, May 7.
5. Wallensteen, P., Nordquist, K-A., Hagelin, VB., Melander, E. (1994) Towards a Security Community in the Baltic Region. Patterns of Peace and Conflict, Uppsala University, Uppsala.

THE RESTRUCTURISATION OF ESTONIAN INDUSTRY

MARGUS LEIVO

Until the present, all governments of the Republic of Estonia followed a relatively identical, very conservative policy. In essence this resulted in the sudden implementation of liberalisation and macro-stabilisation which could be termed shock therapy. Such a very conservative economic policy without clear and precise formulation of ideological foundation was implemented simply because Estonia did not have another choice.

The structure of industry in Estonia did not meet the requirements of the revitalised Estonia, which is in the process of making the transition from an unstable economy to market economy.

Because of the very large market of the formerly closed system of the Soviet Union, the large quantity of natural resources and deficits in the economy and industry did not have to keep up with competition on the world market. The principle difference of the price formation system (the especially low prices of energy and resources) allowed for the development in Estonia of industrial branches which required much energy and resources but produced poor quality goods.

Industrial enterprises in Estonia, which previously exported goods to the Eastern market, according to the accepted plan in Moscow, fell into a completely new situation. Where earlier enterprise management solved mainly industrial-technical and supply problems, now there are new issues of what and for what market to produce.

After the monetary reform conducted on July 17, 1992, which separated Estonia from the rouble zone, the situation became even more complex. The result of the monetary reform was still noticeable up to its implementation. In expecting monetary reform, a majority of enterprises no longer wanted to sell their production for roubles and stored their production mainly in warehouses.

After the monetary reform, financial operations from the countries in the rouble zone grew very complex. The rapid inflation and unprofitable exchange rate "devoured" the funds of enterprises. The devaluation of the crone increased the price of imported raw materials, due to which firms generated losses.

Many of the economic processes changed and grew so much more complex that it was difficult to separate causes from effects. Unfortunately, industry in Estonia had to start all over again from zero. This was not our goal. Despite our efforts to avoid this and lessen the burden, in many cases it was necessary to start

K. Prunskienė and E. Altvater (eds.), Transformation, Co-operation, and Conversion, 35-40.

with the elimination of old structures, and only after this begin to build them anew. We understood that no matter how unfortunate it may be, industry in Estonia would not be able to be reformed gently and unpainfully.

And so it proceeded. Five years in a row, industrial production declined compared with earlier years. In 1991, industry comprised 34.3% of the GDP in Estonia, and in 1994, this was only 18.6%. Of the main production in 1989, only about 45% remained.

If someone wishes to disagree or confirm that structures of industry in Estonia could have been changed according to a set plan, simply and unpainfully, then I would ask: "Why did this not work in any socialist country?"

To affirm that in all former socialist countries incompetent individuals formed and implemented industrial policy, would contradict the theory of probability. It would be more correct to come to the conclusion that events had to take the turn they did and that other possibilities were simply non-existent.

The open economic policy in Estonia is not simply a craving, but an objective necessity. With its very small internal market the country simply cannot exist in a closed system. It was forced to accept an open market position. And because the economy of Estonia is absolutely open to the external market, development of industry in Estonia depends on both internal factors and on several external factors as well.

Mostly, we are dependent on the success of external economic partners. We depend on the development of the German economy to ensure the strengthening of the sturdy exchange rate of the EEK. For us it is mandatory to find a necessary equilibrium between these two currencies.

I will now briefly describe a few changes in the main principles of industrial policy in Estonia during the period of structural changes and integration of Estonia into European processes. I will make the assumption that the goal of Estonia is to enter the European Union.

We cannot under the largest desires remain between Russia, which dreams of re-establishing its empire, and the European Union. After our application to the EU, we await political security and a stable development of economy.

What were the principles of industrial policy during the period of structural change and integration to the European processes?

1. Industrial Policy Should be Liberal and without Protectionism.

The basic inadequacy of the protectionism factor lies in its freezing of the existing irrational structures of industry and the insufficient competitive capabilities. For this reason protectionist methods should be implemented only when there is something to protect.

During the period of re-establishment of independence, such companies did not exist in Estonian industry. In 1995, there are still none. The selective assistance of industrial sectors, companies or regions is very threatening for two reasons:

1. In the complete transition of the economic environment, it cannot be said what industrial structure is ideal, which spheres of activity should dominate in Estonia, and what will be the costs of the possible subsidies for these sectors' development. Because of limited information and the multitude of unknown factors, prospective implementation remains rather problematic.

Mistakes made in setting privileges will bring irrevocable economic and social losses. For this reason it is more reliable at the given stage to direct development toward market regulating processes.

2. The selection of industrial branches, enterprises or regions in order to provide them with subsidies or protect them, in essence appears as directing taxpayers' money in one or another region of economics. Such an industrial policy raises comparatively high requirements in order to reach social and political agreements between various social groups, political parties and regions.

Differences of opinion between producers and consumers could arise. The unemployed and those who think that unemployment does not affect them, react differently toward protectionist industrial policy.

The Ministry of Economics in principle does not deny the identification of priorities in industry. We bear responsibility that under the conditions of inadequate information and uncertainties, the poor selection priorities and its support are a bit worse than market regulating processes.

With the current situation, industrial policy in Estonia should in essence trust the decisions (external) of investors and that the financial capital of commercial banks will concentrate where it is more profitable and reliable.

Prospective branches of industry or separate industrial enterprises in Estonia determine its ability to find a niche in the world market, to obtain mandatory investments and credits, successfully develop these branches, etc.

The market chooses its winners. The task for Estonian industrial policy is to create necessary laws which provide for the activities of market mechanisms, development of infrastructure and competition.

2. The Main Part of Industry in Estonia Requires Privatisation.

And this principle does not result from some abstractly correct ideology but from the arisen situation.

As I stressed in the beginning of this paper, the structure of industry in the transitional period from a planned economy to a market economy did not correspond to the needs of the re-established independent Estonian Republic. This structure should have been changed.

Very large investments are necessary for the structural change in Estonian industry. The production capacities increasingly depend on technological processes and utilised machinery and equipment. New technologies and equipment are very expensive. On utilising cheaper technology and creating cheaper employment positions there is no discussion. In the modern world market they cannot compete.

The utilisation of government budgetary funds in Estonia for structural-political goals is very limited. The offered Western economic assistance and loans is also insufficient. Besides, the government, as a rule, is a poor host and is simply not in a position to deal with such a large number of enterprises. Namely for this reason, it was decided to privatise a large part of enterprises. Private capital must work on its own.

3. It is Mandatory to Completely Co-ordinate the Flow of External Capital into Estonian Industry.

Applying only aggregate internal savings toward the re-organisation of industry in Estonia, this would take ten years. This road of development is completely unrealistic, since history would not support such slow development. Since local capital is insufficient, rapid reconstruction of industry is possible only with adequate external investments.

The flow of foreign capital to Estonia would provide local enterprises with both single orders and long-term and stable possibilities, it would create new jobs, it would increase the government tax base, etc. This would "push" the Estonian economy on the whole.

4. It is Necessary to Depend on Individual Entrepreneurs and Persons.

In Estonia, no classical industrial sectors exist, and it is not possible to create them in the near future. This is due to, on the one hand, the small size of Estonia, and on the other hand, the industrial policy of Moscow, which in its time completely limited the appearance of independent industrial complexes in the republics of the former USSR.

In such a small republic as Estonia, with an open external economy, it seems to be impossible to create a classical branch satisfying all mandatory requirements. Because the utilisation of funds from the government budget for structural-political goals is quite limited, and industrial policy hopes for correct (external) solutions from investors and commercial banks, industry in Estonia realistically should be unavoidably oriented towards individual enterprises.

As can be observed in many industrial enterprises which in their sectors are the only ones, raw materials are basically imported, and a large part of finished production is exported: co-operational ties with other enterprises are practically non-existent.

Industrial enterprises in Estonia could find their place in international markets. The Ministry of Economics bears responsibility for the fact that the rentability of assembly work is usually lower than the manufacturing of finished production. In some instances, such activity could determine the fall of production potential in enterprises.

If an order is placed for both technology, equipment and materials, then the necessity for constructors and technologists is practically eliminated. Universal, wide profile enterprises could become fully dependent on Western firms in sub-production. But in economics, as in life, there usually is no choice between good and bad possibilities. In essence the choice exists between a few poor ones.

Since the majority of industrial enterprises do not possess:

- international recognition of trademarks,
- reliable market analyses,
- market ties,
- know-how,
- capital for scientific research and implementation,
- capital for modern marketing, or
- capital for risk with the possibility of failure,

sub-contract work for western firms appears to be one of the roads out of the difficult situation. Such activity provides possibilities for supplying equipment, reorientation to western markets, in many cases to increase industrial-technical levels of production and improve the work culture.

5. It is Necessary to Completely Support Small Business.

In other countries, there are few engineers and managers with higher education in small businesses, since in large firms the salaries are typically higher. In Estonia the situation is currently completely the opposite: as a result of the structural crisis, small businesses are headed by many individuals, formerly large government engineers and enterprise managers, who see further prospectives for the development and expansion of industry.

6. It is Mandatory to Introduce a System of Eurostandards in the Near Future.

If we wish to successfully integrate into European processes and export our production, then we must hurry. The EU-issued so-called White Book provides instructive recommendations for conducting this work.

As a result of the conducted industrial policy and thanks to the constant development of industry under new economic and political conditions, currently industry in Estonia is stabilising. With this it is not worth dreaming about the re-establishment of the stabilisation of the earlier industrial volumes (capacity), but to keep in mind the formation of industrial structures satisfying the requirements of the Estonian Republic.

A majority of industrial branches at present have already begun their economic growth. This shows that although certain convictions of ours on some issues of industrial policy could appear naive, in essence our chosen path was still correct.

Perhaps now the stage of development is truly approaching, when it is time to deny the primitive expression that "all developments deserve simple explanations and all problems deserve simple solutions." Perhaps now it is truly time, along with the teachings of Milton Friedman, to read the book by Paul Krugman on protectionism and competition in modern economics.

The key role in this work undoubtedly is played by the Ministry of Economics, which is concerned with the strategies of economic policy in Estonia. Our sphere of activity should be above all directed toward the creation of a support system of economic policy and economic strategies, support of the export-quality production of businesses, etc. Our task, with the help of statistical analysis and research, explains where the potential of Estonia lies with respect to both industrial branches and the regional aspect.

CONVERSION PROBLEMS OF THE POLISH DEFENCE INDUSTRY

W.M. GRUDZEWSKI AND I. HEJDUK

1. Introduction

The restructurisation of the Polish defence sector is very much dependent on the War Doctrine. Poland is currently interested in entering NATO. Therefore, all of our priorities in converting the Polish Defence industry are structured around our future co-operation with NATO.

The acknowledgement of "partner" status in correlation with the European Community created a bright future for the possibility of working with NATO. Currently the cores of the Polish government, responsible for the military and defence sectors, have based their co-operation on the political and military structures handed down by NATO.

It is necessary to mention that the equipment used by the Polish defence sector is far from European standards. In fact, most of the equipment at our disposition is based on the old generation and is not only physically inadequate but also economically obsolete. The fast process of political and economic changes, and mainly the ones reached in the economy through privatisation, restructurisation of most of the heavy industry, machinery and chemical industry, have called for further changes in the defence sector so as to satisfy the needs and wants of the country. This goal may only be attained with the help and active communication between the public and private sectors of the Polish Republic. All this must be accomplished on a smaller scale and be firmly based on economic laws, balancing supply and demand. The effectiveness of this task is primarily dependent on the funds relinquished from the already scarce budget for the purchase of new and basic equipment, which should represent complete reliability, high quality and international standards.

The end of the cold war has slowed down the military sector, and, therefore, the supply of military equipment has declined drastically. The future demand of these items will greatly depend on whether Poland becomes a member of NATO and on the capabilities of the Polish industry to produce equipment up to par with European norms. We also must take into account that Poland cannot guarantee high technology, neither now nor in the future, which is the root of all military equipment. This type of equipment must be imported. This fact poses a great threat to the Polish defence sector. Most of these industries will have to undergo a complicated change-over from the military to a public sector. The

K. Prunskienė and E. Altvater (eds.), Transformation, Co-operation, and Conversion, 41-46.

"change-over time" is very important due to the fact that in the case of conflict, Poland may satisfy only 40% of its military needs. To this day, this slowly dying industry, which once operated in alliance with the Warsaw Pact, is now kept alive by supplying equipment to the Ministry of Defence and the Ministry of Internal Affairs, which utilise 20% of this sector's potential. On top of all this, the chances of export with high competition, the embargoes imposed on certain countries and the "negative letters," all pose a burden and a great threat on the process of converting this industry.

2. The Characteristics of the Defence Industry

In many countries, the defence industry sector is placed as a priority in the economy. A direct effect of placing defence at the top of the list is the birth of new technology, better materials, lower costs and highly trained personnel which in turn stimulate progress in many other fields. Many technical procedures and their solutions as well as high quality, accelerate and evolve the quality and effectiveness of production. Currently, there are unfortunately many means of production lying idle, which need to be transformed to answer the demands of the public sector while again obeying domestic and international standards.

The Polish defence sector, composed of 31 institutions, has already, to a great extent, undergone changes and has been strictly regulated by the newly established Trade Laws. In other words, these laws forced each institution to balance their costs and income, forcing them to be more profit-oriented. Many such industries have found themselves in a very difficult financial situation. Most are non-profit with enormous debts, are over-employed, do not utilise their potential man-power, have sunken costs, and use old technology, but more importantly, they impose high deficit costs on the nation. Most attempts to bring these institutions out of debt have brought very few results. We fear that many of them will fall into even greater debt than before. Regardless of these problems, this sector may be revitalised with properly implemented restructurisation programmes. It is our opinion that institutions which are undergoing structural and reformational changes should place their highest concerns on changing the employment structure, organisation and an "imperative" change in their line of production. This restructurisation should focus on changing the military sector of production to a public sector. All of the programmes and conversion methods suggested have to be feasible and of course financially stable, and they must be based on extensive market research, taking into account the competition aspect, feasibility and asset liquidity. In short, in order to carry out such a difficult undertaking, a company business plan must be drawn up and a thorough feasibility study completed.

It is necessary to reduce negative trends in the development of the Polish defence sector.

3. Conversion Goals and the Restructurisation Programme in Poland

During time of peace, 60% of the Polish military industry should be directed towards the public sector. In China, it is mandatory for all military-based industries to direct 50% of their output to the public sector.

Conversion policies require special care due to the size of the Polish industry, which could direct its output, based on the guidelines dictated by the laws of supply and demand and defence requirements, on substituting military goods with electric tools, lathes and items of use to the farming sector. Therefore, a company producing tanks and heavy armoured vehicles could change-over their production line to produce heavy building equipment, soil tillers, technical help and support in building roads and highways, or create tools to make city cleaning and sanitation more effective and efficient. In another case, an institution which used to produce tank missiles, could now manufacture cement mixers, propane tanks of equipment for small butcher shops. These are some of the many feasible plans which could be implemented in Poland. Of course, all proposed plans of conversion will have to be checked and will have to direct the company output to a given field in the public sector within the interests of foreign and direct investors and the needs of the private sector.

The conversion process requires integration of unused manpower, after sustaining all activities in the institution. This must be done while looking at modernisation possibilities and estimating the needed capital to produce for the public sector. All this is connected with the amount of funds required to undertake such a structural change and at the same time balance its sources. The financial needs for these types of conversions are far beyond the capabilities of the Polish banks in the form of low interest loans or government stocks and bonds. These companies will not be able to overcome their financial difficulties without the help of foreign investors and therefore may lose their chance for conversion.

Another very relevant problem is the availability of technology and know-how, which allow new goods to be produced on a competitive level here and abroad. Conversion requires a new tax system, loan preferences, donation and subvention in each company in such a way, so as to enable each company to step over into the profit margin. There will also be a need to come up with new organisational plans and increasing export capabilities. On the one hand, this requires self-initiated exports, and on the other hand, the freedom to work with any domestic and foreign distributors of the produced goods.

4. The Needs and Limits of Restructurisation

Every company undergoing the restructurisation process will have to establish:

- a privatisation programme (change-over plan);
- a proposal for a new organisational system;
- restructurisation of its capital and assets;

- restructurisation of the labour force;
- a production profile;
- proposed regulations on the functioning of the company;
- proposed means of funds through domestic and foreign investors;
- a list of strategic foreign or domestic investors.

In many cases, in order to solve the problems of larger companies in the military field of production, it will be necessary to get rid of the "ill" systems such as "holding." It would also be wise to prepare a business profile revealing their financial state, their place in the market (both foreign and domestic), the state of their production files, the short and long term restructurisation plans and stating their potential foreign or domestic investors. This kind of research will utilise our experts and sometimes even foreign experts.

5. A Practical Example of Conversion in Poland

A primary example could be a company in which 97% of production was targeted at the military sector prior to its restructurisation. After a detailed analysis of the company and its labour capabilities, the programme was targeted at the public sector by providing cash registers for the growing field of small business and private shops as well as chain stores and restaurants. A regular price also needs to account for the Value Added Tax (VAT). In order to quickly and successfully implement this project, a foreign partner is a must, someone who could deliver the proper technology and know-how and also secure funds for this purpose.

This allowed for an increase in production of up to 70%, an increase of production capacity by 25% and a significant rise in employment by about 40%. There was an increase in pay and finally profit for the entire company.
Prior to this change the situation was critical.

Besides the restructurisation of production, the company entails a change in the organisational system as well as the introduction of "lean management." This creates a slimmer organisational structure while still fulfilling their duties and goals.

Another field of restructurisation is in the field of employment, which as a result allows for a change in the structure of employment, removing the unnecessary positions and increasing employee effectiveness.

Organisational changes are tied to the distribution of physical assets and liquid assets. This is facilitated by determining profit centres within the company. These changes are driven by the future structure of the holding company and preparation activities in the privatisation and sales of public shares.

In a team effort with German companies COGNOS and FRITZ WERNER, we undertook the restructurisation of a military-oriented institution producing goods for the needs of the Warsaw Pact. The dismissal of the Pact, politics and the unthreatening situation of Poland's neighbours all caused a fully new production plan to be drawn up in these institutions.

The point of entry was the need to improve the current plan of production. As a result of market research conducted by both Polish and German specialists, a plan was established to change production to cooling aggregates, lathes and interchangeable parts for the foreign partners. Other parts of restructurisation were in waste management and environmental safety, rebuilding of most company machines and vehicles and installing a desulphuriser.

5.1. PROPOSED CHANGES FOLLOWING THE CONDUCTED ANALYSES

- A complete alteration of technology and the methods of production used in construction work and planning and management productions;
- The implementation of a new organisational structure "holding system," in which a company was divided into smaller parts, each self-financing.
- The improvement of the functions of the new structure through the implementation of computer-assisted management.

5.2. RESULTING CHANGES

- The obtainment of additional cash flows and financial results;
- An increase in production efficiency from 30% to 60%;
- Reduction in production costs in the first stages of implementation by 15%;
- the implementation of a new concept of Total Quality Management or TQM;
- A doubling of average returns;
- The elimination of all unnecessary installations of production;
- Renting or selling unused capital;
- A drop in employment by 20%;
- The implementation of a motivational pay system;
- An increase in capital per person.

Moreover, a marketing system was set up within the project, oriented towards client and company benefits. A market strategy was also established based on production and distribution. Thanks to all collected data and systematic observations, the competitor was able to obtain additional clients and a new entry into the market. This project also increased net worth sales.

Yet another good example is the financial restructurisation of large contemporary companies in the military field. Due to the recession in the country in the 90s, a drastic decline in demand from MON and MSW and budget cuts caused the companies to go bankrupt. After implementing changes, the companies focused on a totally different field of production. They quickly began producing goods for the public sector, which had a high demand among the consumers in the market. The cost of restructurisation was great, and only a portion of the funds came from the State budget. This was the simplest way of getting into the debt hole, loose liquidity value and credit capabilities. The capital was finally provided as high rate loans from banks.

In such a situation, the only reasonable solution would be the restructuring of a company leading to the elimination of debt. This requires specific agreements with creditors. Such restructurisation should result in the profitability improvement of a company.

6. Conclusion

Each undertaking of such programmes in Poland should produce the following results:

- A full change in the assortment of production in at least 50 companies, which were previously oriented toward production for the military sector;
- A partial change-over in the remaining 30 companies;
- The implementation of rules and basic procedures limiting military production in the military sector from 30% to 40%;
- The privatisation of capital sources, assets, finances, organisations and the restructurisation of employment;
- The obtainment of new foreign investors and partners;
- The Commercialisation of all institutions in the Military Sector;
- The privatisation of most of the old military sector producers;
- The guarantee of low rate loans from Polish as well as foreign sources;
- The obtainment of loan guarantees, for foreign investors, of 2-3 billion old zlotys;
- The temporary dismissal of debts to companies with a bright outlook after conversion;
- Subsidies to assist in the implementation of these programmes, connected with new technology and solving organisational problems.

MECHANISMS AND PERSPECTIVES FOR THE CONVERSION OF THE POLISH ARMS INDUSTRY

KATARZYNA ZUKROWSKA

1. Introduction

During the systemic transformation, the Polish defence industry was the hardest hit sector of the economy, resulting in its large scale downsizing and in difficult adjustments. The remaining capacities still seem too large for the demand of two markets: the Polish and the international market. Some of the enterprises indicate recovery in their current financial liquidity, which does not mean that financial constrains have been overcome. This paper will discuss: (1) what has been done in the Polish defence industry until now, (2) what main problems the Polish arms industry now faces, and (3) how existing problems can be overcome.

2. Changes in the Defence Sector

There are generally three options in choosing a strategy for downsizing the defence industry: (1) the paternalistic approach; (2) market solution; (3) and the mixed approach. In the first strategy the government attempts to cushion the negative results of adjustments, which in practise prolongs the adjustments, as enterprises count on the intervention from the political centre when they experience difficulties. In this respect, management is not forced to utilise the best possible and most effective solutions from the economic point of view. This in turn causes the prolongation of the transition period. Market solutions are often unexpected and brutal for the industry, but these force management to seek proper strategies for enterprise survival. The social costs of market adjustments are comparable with the paternalistic approach, as in both cases money has to be spent from the budget (a) on intervention and (b) on allowances for the unemployed. Employment reductions are similar in both cases, no matter what arguments are used in support of paternalistic measures. The mixed approach prolongs the transitional period in contrast to the market solution, which increases the costs of the process. This in fact contradicts the requirement of returning to stable financial footing for the enterprise in as short a time as possible. With the extension of the transition period, the costs of transition are not only increased but the enterprise also loses its customers and its share in the market.

K. Prunskienė and E. Altvater (eds.), Transformation, Co-operation, and Conversion, 47-57.

Poland applied a market-driven strategy of downsizing. This was painful but at the same time rather effective, but this does not mean that the problem has been solved once and for all[1.] Recent worker strikes in the arms industry indicate that tensions still exist in this sector, and production capacities still exceed total demand. Poland is attempting to introduce solutions on its own in the defence sector. National solutions in practice mean the continuation of self-sufficiency in arms production and a limited private share in the sector, especially where foreign capital is concerned.

Market-driven strategy is not univocal with total lack of intervention from the government, as such a solution would be utopian in the case of the arms industry and other branches of heavy industry, especially under the conditions of geographic concentration and the high regional unemployment rate. But intervention in this specific sphere has been treated as an ultimate too, forcing invention from management, which replaced ties with the centre by horizontal relations. Enterprises had to renegotiate the conditions of their credits and debts with banks and other enterprises. This led to new interdependencies and a definition of common interests. In drastic situations, when enterprises were completely paralysed by the lack of financial liquidity, part of their debt was reduced.

The level of indebtedness, especially to the state social insurance company (ZUS), still remains high despite a partial debt reduction by the Ordinance of the Ministry of Finance No MP 10. Arms producing enterprises may be divided into three categories - as far as profitability, fiscal liquidity and assets structure are concerned: (1) group of economic security - 9 enterprises; (2) group of moderate risk - 13 enterprises; (3) group of high risk and threat of bankruptcy - 9 enterprises.

According to the statement by the secretary of state in the Ministry of Industry and Trade, "the repayment of debts will be concentrated on the period 1995-1996, as a result of conciliatory and composition proceedings. There is a real danger of renewing the indebtedness of enterprises if no serious preventive measures are taken.[2]"

The idea behind a state policy towards the arms industry was meant to introduce long-lasting and homogeneous systemic solutions that created a basis for the future economy of the country, embracing also the arms industry, which must be closer integrated with the civilian economy[3.] In such conditions the policy towards arms industry consisted of two elements:

1. a set of general steps covering the whole economy;
2. particular moves, addressed to the arms industry sector.

State activities in the marketization process of the whole economy embraced: price liberalisation and foreign exchange; macro-economic stabilisation; institutional changes within national borders and outside the country. Marketization of the economy also embraces the defence industry - an inseparable part of the economy. The defence industry cannot function apart from some general rules of the market game that are introduced in the process of marketization. The weight of this problem in the whole transformation process depends on the scale of the military sector and its share in the total industrial production of the country.

Government activities towards the defence industry in Poland covered: (1) budget policy (reduction in the defence budget, univocal with reduction in demand as well as with the cancellation of subsidies); (2) legal regulations (the creation of an institutional framework for the future functioning of the industry)[4;] (3) export supporting policy (co-financing exhibitions); (4) institutionalisation of external contacts, (5) concentration of production, (6) debt reduction; (7) commercialisation; and (8) compensation of part of the losses[5.]

The restraint of the state policy towards the arms industry was dictated by two conditions: limited financial resources resulting from stabilisation policy and considerations that any helpful hand given to enterprises will limit their own invention of finding proper solutions for survival. In new economic realities market enterprises seek partners, forming chains of elastic specialisation alliances. Such a solution finds its path under Polish conditions as well, and this is not simply limited to a "copy-cat" pattern artificially implanted into the transitional period, but a formula invented by those struggling to survive.

3. Effects of the Acquired Strategy

According to the Ministry of Defence (MoD), military expenditures in 1991 (constant prices from 1994) have reached a level of 52,291 billion zloty, comprising 7.5 per cent of the state's budget and 2.0 per cent of GNP. In 1994 the expenditures, in 1994 constant prices, were 47.855 billion zloty, 7.3 per cent of the budget and 2.5 of GNP. In 1995 this decline continued, with a comparative share of military expenditures in the state budget reaching 6.5 per cent and 2.2 of GNP6. The decreasing trend was directly affected by: (1) the policy of "difficult money", (2) the downsizing of military expenditures related to the end of the "Cold War". This means a 56.5 per cent drop, measured in real terms, in comparison with 1986 levels. This was also followed by a serious restructuring of expenditures leading to an increase in personal expenditures and a reduction in purchases (by 50 per cent) as well as in R&D (research and development).

A system of legal regulations was created, that embraced:

- regulations of the ownership changes in the arms industry;
- regulations of transfers of advanced technologies;
- terms for granting concessions and control over arms trade;
- the creation of an organisational and structural model of the defence industry.

Because of the adjustments, the number of the companies in the arms production decreased from 90 to 44. 31 of this number are supervised by the Ministry of Industry and Trade, while 13 are supervised by the Ministry of Defence (MoD). In practice this means that only 31 enterprises are engaged in military production, while the remaining 13 specialise in repair works. Employment in the sector declined from 180 000 to 40 000. In 1994 the share of special production in the GNP was 0.5 per cent, which reflects a 1 per cent drop from the 1988 level (1.5 per cent). In 1993, the value of sold production and services (special and civilian) of the defence industry slightly exceeded 1 per cent of the total sale value in domestic industry. This proportion was unchanged in 1994.

The reductions of MILEX (military expenditures) did not influence the state of self-sufficiency of the Polish defence industry as compared to the demand of MoD. After the reductions, Poland is still self-sufficient by 40-80 percent. The 40 percent difference reflects the lack of compatibility of the Polish equipment towards the NATO requirements. This gap may be easily covered by closer international co-operation both in production and R&D. Such solutions require political decisions on both sides, i.e. Western and Eastern partners with Poland. Moreover, they are preconditioned by the decisions on the eastward expansion of NATO.

Enterprises which form a basic group of the defence sector in Poland co-operate with 353 supporters, 65 of which are of particular importance to defence production[7].

Companies with a low share of military production in overall output experienced fewer problems in moving to the civilian market. Despite these difficulties they have approached by shrinking demand on the home market and increasing competition abroad. This state of affairs may be considered a mobilising factor, forcing companies to adjust to new rules of the game.

The restructuring of the Polish arms industry was based on two types of decisions:

1. market selection, i.e. selection of winners and losers, eliminating those who lose and upgrading those who win;
2. political and strategic decisions, i.e. taking into account the meaning of this specific type of production for the country's defence capacities.

Enterprises engaged in the defence production are currently aiming at achieving three goals in improving their financial standing:

1. changing the proportion between the civilian and military share of production, aiming at a 70:30 ratio;
2. increasing export sales by 15-20 percent;
3. reorganising the structure of the arms industry into 4 branch holdings:
   - classical weaponry producers, which would embrace producers of ammunition, missiles and explosives. This would include 14 enterprises, employing about 32 thousand workers[8];
   - aircraft producers, embracing 5 "PZL" enterprises, employing about 24 thousand workers;
   - producers of radiolocation equipment, optoelectronics, which will consist of 5 enterprises, employing about 10 thousand people;
   - armoured vehicles producers, consisting of 5 enterprises, employing 22 thousand people.

The first stage of the process of ownership avoids its termination. State owned companies are being transformed into one-holder joint stock companies owned by the State Treasury. The future ownership structure of the stock package foresees that the control package is left in the hands of the state (51 percent), shares of the workers employed in the companies will amount to 20 percent, while the remaining 29 percent of shares will be sold out on the home market.

A commercial approach towards arms trade can be seen, occurring simultaneously with searching for co-operants abroad. These new tendencies may be exemplified by several facts. For the first time Polish arms exports are presented in a catalogue, which provides all necessary information regarding quality of the equipment and its technical features. Two international arms fairs organised in Poland serve the same purpose. Both will be taking place in September, the first one to be held in Kielce (offering a full range of Polish arms and military equipment), the second one in Gdansk (offers so called small weaponry and quartermaster's equipment). International co-operation agreements on production are signed with companies coming from such countries as Germany, Great Britain, France, Canada, Belgium, Spain, Austria, Switzerland, and Finland.

Commercialisation and the shift towards civilian production is seen in industrial production as well as in R&D activities. The first tendency can be illustrated with an example from the Warsaw Optical and Electronic Enterprise, which supplies the market with high-tech electronic cash-registers produced in co-operation with a German company. The second tendency is reflected in the activities of the Military Technical Academy, which uses most of its scientific capacities for civilian research, resulting in numerous international rewards, mainly received in Brussels on International Technical Fairs.

Changes in the arms producing sector R&D as well as its training lead us to several questions:

- When will political decisions be undertaken to bring this sector closer to international co-operation?
- How may existing capacities in R&D be promoted and utilised without political suspicions that such co-operation is counterproductive to national interests?
- Under what conditions will co-operation among Lithuania-Poland-Germany in arms production be acceptable?
- Will it be possible to utilise the existing capacities in training medical specialists from Lithuania, Latvia, Estonia and the other former republics of the Soviet Union in Poland? Can this kind of education embrace other post-communist countries? Can Western specialists participate in such programs as well?

These and other questions of similar type are not yet answered, but it seems that their formulation is progress in itself, and in a short while we will know answers to such questions as: (1) Are we in favour of such solutions or are we against them? (2) What will each type of solution bring into the practice of international relations? (3) What has to be done in order to support such ideas? (4) What are the conditions for their implementation? (5) What impact will such solutions have on confidence-building measures? (6) How will they influence the stability of the international security architecture?

4. Adjustments as a Permanent Feature of Market Economy

Economics is a process. In Latin we would say PANTA REI - Everything is changing... Western governments and Western enterprises are used to this, as they have always been living in a state of flux, while countries in transition are not only learning the new rules of the game but must also find their new places in the new environment and under new conditions. In other words, they must learn how to read market signals and how to react to them, so that decisions would lead them towards new markets, new production, new technology. Is this possible, and if so, how quickly?

The Secretary of State in the Ministry of Trade and Industry indicates the four stages undergone by the arms producing industry since 1989. These stages are as follows:

1. The first stage - the years between 1989-1991: diversification of production and attempts to reduce costs. Limited knowledge of market rules and of management and the transitional collapse of the market drove enterprises into debts. The Ministry has prepared a programme of restructuring which was never put into practice;
2. The second stage - the year 1992: the Ministry prepared two versions for defence industry restructuring. Finally the budgetary version was chosen, but this was also never put into practice. This stage covered the following moves which were undertaken on the enterprise level: (1) concentration of production; (2) reconversion of production; (3) changes in organisational structures; (4) implementation of individual restructuring programmes;
3. The third stage - the years 1993-1994, covered some moves which were achieved on the enterprise level within the framework of commercialisation and debt reduction of the enterprises, prepared and approved by the Economic Defence Council, at its meeting in August, 1993;
4. The fourth stage - the years 1995-2010. According to the IDA-EC PHARE Contract No 223, the preparation of the defence industry restructuring programme for 1995-2010 is under way. In December 1994, the first stage was completed with the preparation of the report on the sector's position. The restructuring programme is guided by four principles: (1) a definition of the needs of the security system; (2) providing the framework of restructuring; (3) working out the implementation strategy; (4) setting proportions between social demands and competitive requirements.

All of my interviews with managers of arms producing industry enterprises indicate that the sector would appreciate some specific treatment from the government, which, according to them, should embrace the possibilities of receiving privileges in form of low interest credits, support for export promotion, etc. The main question here concerns the fact that lobbies - if strong enough - may gain some privileges which would in turn mobilise other interest groups to demand at least similar treatment. A departure from the strict rule of limited intervention produces a snow-ball effect which is difficult to stop. Moreover, practice shows that when the state is forced to introduce such measures, it faces great difficulties to withdraw them when they are no longer needed. In such circumstances it would be better, having the diagnosis of the sector, which seems to be the easiest part of the job, to create an independent body, with a group of experts which would act on a case-to-case basis in advising the moves but not representing the government and the budget money. This institution could serve as a co-ordinating body of all the moves taken towards the defence industry: indicating what legal regulations are needed, how the enterprises should act in the new conditions; to whom they should turn for money; how to survive restructuring and repay debts at the same time?

The second part of the solution concerns the question of legal framework, within which the future defence sector could operate. This part of the restructuring programme must be solved completely by the government. Here the representatives of the government/parliament have to answer crucial questions on ownership, scale of engagement of foreign capital, structure of production, etc... Delays in answering these questions leave the decisions about the sector totally in the hands of the market, which may be conducive to further reduction in number of the enterprises engaged in arms deliveries.

In my opinion, the military sector can only survive under the condition of international co-operation. Thus such a solution requires political decisions, and this is out of question... Moreover, too much paternalism will limit the incentive to change, to adjust, to economise. At the same time, a specialised institution which is able to give effective advice for a reasonable price could be one of the best solutions for the sector. Military production is lucrative and there are many reasons behind it. Under such conditions, the institution in question must not only be trusted in Poland but also abroad in order to effectively seek counterpartners. It must possess the required contacts at home and abroad.

The need for such a body can be proved by showing the development of a special economic zone in Mielec[9], a town which specialised in aircraft production and currently faces deep structural crises with all the consequences. The major negative effect concerns the fall in employment from 20 500 in 1988 to 6 750 in 1994[10]. The above proposed institution could advise ways of attracting foreign capital and promotion of the activities carried out[11]. The main advantages of Mielec are as follows: an airport; its close distance to four borders; its close distance to a future European highway; exemption from taxes; accelerated amortisation of capital; limited local taxation; limited requirements to start business; government guarantees that listed privileges will last for the planned period.

The strategy that was utilised in the case of Mielec relates to effects in other parts of the world. It is enough to mention Europe, North, Central and South America, the Far and Middle East. In the US alone there about 200 Special Economic Zones. They are also established in some post-communist countries, like Hungary and Russia.

The Special Economic Zone (SSE) in Mielec encompasses 575 ha grounds of the former WSK PZL (including the 350 ha airport). The zone consists of companies created as a result of the capital restructuring of the former state factory (12 limited liability companies) and other smaller, newly created private companies. The 150 ha to be used by new investors include ready to use buildings for production, warehouses and offices with a total area of 80 000 sq. m. The infrastructure is relatively well developed with telecommunication networks, electricity, heating systems, compressed air, internal roads, side-tracks, garbage dumps, etc...

Mielec is quickly becoming one of the most attractive places in Poland for investment. The availability of a highly qualified work force is a main attribute of SSE Mielec. It encompasses a wide range of professions and specializations, which result from the aeronautic and automotive engineering heritage. The engineering force has professional experience in such fields as construction, technology, product assembly and sub-assembly.

Mielec creates an opportunity for investors to hire highly qualified workers easily and at a relatively low cost. Such a situation will undoubtedly have a positive influence on production costs for SSE companies1[2]

5. Conclusions: Competitiveness and Competition

Competitiveness of the economy should be understood as its characteristic feature which helps it to adjust to the new continuously changing conditions. Competitiveness of the economy depends on different factors; it is sufficient to mention the most significant of them:

- intensification of competition;
- specific features of the market: size, international ties, level of development, stage of de-industrialisation;
- unity of the market;
- labour costs and productivity;
- industrial traditions;
- exchange rate;
- the interest rate and development of a banking network;
- the development of infrastructure;
- internationalisation of the economy.

All of these mentioned factors also affect the competitiveness in arms production, especially under the conditions of deep changes in this sector in the whole world. Competitiveness of the economy is also closely linked with its stabilisation. As a matter of fact, stabilisation is the common denominator for the marketization of the economy and its competitiveness. The reason behind this is simple. High inflation limits foreign direct investment (FDI), one of the main factors increasing the competitiveness of the economy as a whole. The role of FDI is increasing in the contemporary world. It is still more important in the post-communist world, as countries grouped here badly need investment. They are not able to finance them with their own savings, which are very limited after the stabilisation measures. In such circumstances, foreign savings can come and fill the gap. The inflow of FDI to Poland is on the up-swing, which can be ascribed to best economic results in the region, measured by long and high growth after the quick departure from the transition crises.

Stabilisation of the Polish economy has recently reached the level in which further IMF credits do not require conditionality. Poland was able to repay all its commitments to the IMF, proving at the same time that the acquired strategy has been correctly chosen, and other post-communist countries, still in the pre-stabilisation phase, can make use of it. Especially when bearing in mind that stabilisation is conducive for competitiveness, liberalisation of trade starts to change the branch structure of the economy, GNP begins to grow, inflation is under control, people receive wages.

In its last years of transition, Poland has improved its competitiveness on the international market, surpassing Hungary. The Czech Republic remains the leader in this ranking, losing only in two disciplines to Poland: infrastructure and management1[3.]

The increase in the weight of competitiveness in arms production derives from the following facts:

- priority was usually given to supplies from national producers;
- arms production did not experience similar pressure of costs as civilian production;
- international co-operation in arms production and military R&D was limited, which, as a result, led to the doubling of costs by separate countries;
- standardisation was limited;
- costs were comparatively high when weighed against civilian ones;
- the date of delivery and used technology were of crucial importance for the customer;
- the division of markets into the sectors of East and West also limited competition.

We are currently approaching - let's hope - an era of the end of divisions and closer international co-operation. This should be considered a challenge, an opportunity and an advantage for both groups of countries, those in transition as well as those which are far ahead in their development. Internationalisation and closer co-operation can increase the competitiveness of arms production in both traditional and modern terms. This can be achieved in three stages: (1) the advantage of traditional superiority; (2) inclusion of R&D into the co-operation pattern; (3) a gradual increase in modern factors which determine competitiveness.

The EU has developed a number of tools which stimulate co-operation in this field. They embrace co-operation programmes as well as R&D, flexible specialisation in the region and industrial policy measures. All of these instruments can be utilised in military production, stimulating co-operation as they do in civilian branches. The prospects here are practically unlimited, as according to the Treaty of Rome (art. 223), defence production was left out of competition policy regulations. Now enterprises look for new opportunities in creating their competitive advantages, which help to maintain a stable share in a shrinking military market.

If we take a closer look at the competitive sectors of the economy in countries that are transforming their systems, we will find that some branches of military production are competitive. Moreover, risk analysis is also favourable in case of such production. If additionally this specific sector would have been enriched gradually by an increase of competitiveness of modern factors, then we would see that this sector creates clear opportunities for co-operation - co-operation which is not only meaningful from the purely economic point of view but also matters in the security field.

Warsaw, Nov. 9, 1995 (revision)

References:

---

[1] Henzler, M. (1995) Manewry pod arsenałem. Czy zakłady zbrojeniowe zbankrutują. *Polityka* 20 , 7.
[2] Czerwiński, R. (1995) Issues of Polish Defense and Aviation Industry Restructuring, NACC seminar, 22 May, p. 2.
[3] Pietrewicz, M. (1995) Conditions of Conversion Process in the Polish Economy, NACC seminar, 22 May.
[4] The list of enterprises which require special approval of the Seym (Polish Parlament) while privatizing is given in: Rozporządzenie Rady Ministrów z 26.04.1994, zmieniające rozporządzenie w sprawie określenia wykazu przedsiębiorstw państwowych i spółek o szczególnym znaczeniu dla gospodarki państwa, których przekształcenia własnościowe podlegają szczególnemu trybowi, Dziennik Ustaw, 1994, nr 61, 1159. (Legislation Journal 1994, No 51, 1159).
[5] Zukrowska, K. (1995) Macro-politics of the Government and Arms Industry Sector. Paper prepared for the NACC seminar on Demilitarization and Disarmament in Transition: Socio-economic Consequences, Minsk, Belarus, 22nd-24 March.
[6]. (1995) Wojsko Polskie. Informator 1995, 51-52.
[7] Czerwiński, R. (1995) Issues... paper op. cit.
[8] The figures indicating employment cover both civilian and military production.
[9] (1994) Program of restructuring of the Polish arms industry, prepared by the Ministry of Industry and Trade in cooperation with Agency of Industrial Development and International Development, Ireland, Shannon, 5-9.
[10]. Budzowski, K. Światowiec J. (1995*) Koncepcja Specjalnej Strefy Ekonomicznej w Mielcu, in: Polska w gospodarce światowej*, Sopot, Uniwersytet Gdańki, 30-42.
[11]. Budzowski, K. Światowiec, J. (1995), op.cit.
[12]. (1995) Development Zone Takes Off, *The Warsaw Voice*, 25 September, **2** (Special Supplement Polish Industry Voice).
[13]. (1995) The World Competitiveness Report, IMD, World Economic Forum, Lausanne, 27.

NEW PRIORITIES IN SCIENCES, TECHNOLOGY AND INDUSTRY OF THE REPUBLIC OF BELARUS

MIKHAIL MYASNIKOVICH

In reconstructing its scientific and engineering system, the modern Belarus, like other republics of the former USSR, confronts tremendous problems, much more serious than those encountered by the countries of Central and Eastern Europe. These problems lie not only in the scale and regional diversity, but also in the firmly rooted administrative and command methods of science management. The former state-subordinate economy, which placed its emphasis on the military sector and heavy industry with clearly expressed trends towards extensive development, has created a broad and distorted scientific and engineering basis. In determining the priorities of scientific and engineering development, the state supervised that the intellectual efforts of scientists stayed within the rigid ideological framework. Under such conditions the scientists themselves had little effect on determining the directions of science and engineering development. Despite these severe limitations, the scientists of Belarus, Lithuania, Latvia, Estonia and other countries of the former USSR have outstanding world-level achievements to their credit.

In making the transition to a democracy based on market economics, our countries must transform science and engineering in order for the best in this field to maintain and contribute to their economic, social and cultural revival.

The reduction of state financing and the number of orders and the breaking of traditional ties between republican scientific centres, present a serious strategic task: finding the way in which the scientific and engineering sector should be transformed to adapt it to a market economy and preserve it under the conditions of economic decline.

In any modern economically independent state, the development of science and engineering and the dissemination of progressive technologies are, at the final end, the main factors of general progress. That is why scientific and engineering policy in such states acquires more and more features of flexible governmental initiative strategies, which subordinate structural and investment policies.

The economic and political conditions of the Republic of Belarus, the necessity of faster radical structural reconstruction of the industry, the primitive formation stage of market relations, an objective demand to orient towards intellectual potential as the main resource of development, are serious basics for

K. Prunskienė and E. Altvater (eds.), Transformation, Co-operation, and Conversion, 59-68.

the model of active state scientific and engineering policies and, first of all, for state financial support of scientific and engineering progress. This is connected with the economic crisis in progress, which resulted in the drop of interest of production in innovations and, naturally, brought on a decrease in the number of orders for scientific and engineering products.

Under the transition to a market economy, the future fate of a broad network of research institutions is of special concern. In the past, every branch Ministry had its own scientific and research institutes, design and testing bases. Their real existence corresponded more to the logic of the administrative system, than to the demands of the economy, which resulted in the illogic overgrowth of the system of branch sciences (extra personnel, duplicating and shallow topics of scientific research).

The activities of the branch scientific and research institutes, it is necessary to confess, were often torn away from the requirements of industry. That is why, with the cancellation or reorganisation of some ministries, the branch research and development works began to suffer from a sharp decrease in state financing and from loss of ministerial funds.

At present the institutes and scientific groups try to adapt to new conditions. Reorientation of research and development works, search of new customers and scientific ties will help the most capable scientific collectives to preserve themselves and their potentials under new conditions. Most of the scientific and research institutes which rely upon state donations will have to reorient, unite or shrink sooner or later. They should be ready for this.

The problem of military research and development works looks a bit different, affecting practically all 15 former Soviet Union republics, since in each of them it dominated the scientific sphere.

The best specialists were concentrated in the military scientific and research institutes; they were equipped with the most modern devices and machinery to the maximum extent and consumed more than a half of all expenditures assigned for science development. Such conditions allowed high achievements in research works.

The collapse of the USSR put these scientific institutions in a catastrophic situation (their near closing). Development by the young Eastern European states of their military doctrines, based on the sharp decrease in expenses for armaments, has put before our governments the problem of creating a conversion policy for the military sector of science. Several scientific and research institutes performed parallel civil investigations alongside military orders. These will solve the reorientation tasks much quicker. Some institutes, like those mentioned above, will have to unite and search for customers themselves. Both types may exist in various forms of property (private, state, mixed).

After gaining its independence, Belarus became the owner of several enterprises under control of the former USSR, including big branch institutes previously oriented towards the economy of a great state. Naturally, the industry of Belarus was not able to supply enough orders to them, since most of these scientific centres worked mainly for defence. In order to support them in some way, to give them a chance to reorient towards the needs of the Republic, it was necessary to assign great assets from the Republican budget. This was done partially, but because of the deficit of the state budget it was not possible to satisfy all needs and ensure further development of all scientific and engineering directions. It is still impossible today.

The crisis gave a unique chance to reconstruct science and to make radical changes in it.

From the current science figures, we could expect direct proposals in the direction of scientific progress trends and priorities, perspective planning and reprofiling of some scientific institutes and state support for the most important research.

Under the present conditions, alongside state support, science also requires internal changes -- updating structure, improving the forms and methods of research studies and increasing their efficiency.

The selection of programmes and projects for financing from the state budget should be completed by independent expertise and open contest.

In order to overcome the economic crisis in the science sector and pursue an active policy of innovations, efficient actions by state bodies are needed. Such a special state management body in Belarus, destined to co-ordinate the scientific and innovation activities at the state level, is the Committee on Science and Technologies, founded last year.

At present, active planning of necessary normative and legal documents is in good progress in the scientific and engineering sphere in Belarus; works have been launched regarding the formation of the innovation infrastructure. Upon decision by the government, a programme to develop innovation activities is being compiled and should become a programme of actions for the transient period of adaptation of the scientific and engineering sphere to market conditions.

In the Republic we have begun to restructure the system of financing research and design works (the introduction of a competitive basis and expertise of projects to be included in the Economic Program of the government, the establishment of a Fund for fundamental investigations, providing branch and regional management bodies the right to establish special off-budget funds to finance research and development works), which will assist, in our opinion, in adapting our scientific and engineering potentials to market conditions.

In the process of economic reform implementation, transformations are being undertaken in such important areas as managing research and development works. Although the overwhelming majority of scientific and research institutes and design bureaux continue to operate within the framework of the existing branch structures, these works have been launched in the scientific and engineering sphere to denationalise and privatise and to establish scientific and engineering organisations of non-state property forms. However, today the production share of the non-state scientific and innovation organisations of the total volume of scientific and engineering products is only 3.7%. The situation is practically the same in Russia, the Ukraine and other countries of the Commonwealth of Independent States (CIS). The processes of forming joint-stock organisations in the scientific and engineering spheres and the birth of innovation entrepreneurship in the Republic may be estimated as positive, however, a special study of this problem is necessary to manage the transformation and to put efficient limits on the process of privatising and reprofiling scientific and engineering institutions. Why not make this a topic for an international scientific conference?

Apart from the above, the measures so far undertaken can only give desired results under the condition of a strong state policy based on thorough selection and estimation of the priorities in scientific and engineering development, in distributing state resources in accordance with the priorities chosen and in giving up the false practice of "egalitarianism" in distributing the budget funds among the whole spectrum of scientific and engineering activities.

Under present conditions, the disparity in financing existing priorities with the new socio-political and economic realities objectively dictates the implementation of a new concept of priorities in scientific development within the state scientific and engineering policy of the Republic of Belarus.

The requirements for designing such priorities, which would take into account the weak mineral and raw material supply of the Republic, its difficulties in industry and agriculture, were previously oriented towards the centralised all-Union industrial complex.

The following main principles (concepts) in selecting the above directions have been developed and proposed in Belarus:

- the priority directions designed should primarily address the nearest (transient) period (up to five years), although input into the main list of priorities for a more distant perspective is also possible;
- priorities should ensure the complex usage of the scientific, engineering and production potentials of the Republic and should have a nation-wide and inter-branch, not merely branch, character;
- a solution to the urgent socio-economic problems must be of primary importance;
- scientific and engineering priorities should be present in the main list of directions and should be reflected in scientific and engineering programmes designed in accordance with them;
- the main principles of the directions should be: world level, implemented in the respective international co-operation and acknowledgement; canalisation (generation) of technical and technological progress in Belarus and abroad; an inseparable link to the training of high level experts.

The implementation of programmes in all priority directions should result in an increase of product competitiveness and science-capacious nature, in saving energy and material resources.

Proceeding from the points mentioned above, the following draft list of priority directions for scientific and engineering activities was compiled and submitted for approval to the Supreme Soviet of the Republic of Belarus:

- the development of Belarussian sovereignty -- connected with the development of a scientific basis for forming the economic, scientific, engineering and social policies of the Republic of Belarus as an independent state;
- health protection -- concerning the health care of the population under the most complicated ecological and social conditions, with the aim to preserve the genofund of the state;
- implications of the Chernobyl catastrophe -- the most important direction connected with the change of radiation and the ecological state of the Republic;
- environmental protection -- one of the main problems concerning the inter-relations of the further progress of mankind with the state of the environment, and having the utmost importance for the lives of present and future generations;
- energy and transport -- a direction ensuring the operation of all branches of people's economy;
- production, processing and storage of agricultural products -- a problem directed toward the solution of one of the social problem of supplying high-quality food products to the population;
- informatics and telecommunications -- the basic means of creating and sophisticating an information space;
- resource-saving technologies -- problems directed at increasing the efficiency of using Belarussian resource potentials;
- new materials and technologies -- a direction connected with the decrease in metal consumption in industrial products and the increase in their reliability to world levels;
- improvement of the competitiveness of products in machine-building and radio electronics -- a direction connected with the leading branches of the economy and science in Belarus, where most of the productive population of the state are busy (with the account of conversion).

The concentration of financial assets within the priorities of scientific and engineering development provides the possibility to prevent irreversible implications of those negative processes, taking place in the scientific and engineering sphere and in the economy, in this way assisting in strengthening the national economy and sovereignty of Belarus.

In this context it is necessary to note that science is international in its essence, and fundamental science, in particular, has no technology or methodology secrets and cannot be confined to a department or an institution, and in certain cases even to a state; on the contrary, the more publicity, openness and co-ordination is associated with these activities, the more positive are the results of scientific investigations. Fundamental science is unified as is the object of its studies: the phenomenon of nature and society.

Taking all of this into account, the Academy of Sciences of Belarus, which is responsible for the development of the fundamental sciences, has determined the priority directions of fundamental investigations, which are the basis of all branches of public life, as only they feed both science and practice with new ideas and thus provide reliable scientific procurement of the future.

The priority directions of fundamental investigations in the field of physical and mathematics sciences include such branches as laser physics, optics, microelectronics, radio physics, plasma and solid state physics, informatics, mathematics.

In the area of physical and technical problems priorities are -- physics, chemistry and mechanics of surfaces, processes of production and processing of new materials, thermal physics of foundry processes, etc.

Fundamental investigations in chemical and biological sciences should be directed toward solutions to the problems of obtaining new polymer materials, the development of methods of fine organic synthesis, studies of the structure and functions of biopolymers, determination of the physiology, biochemistry and genetic bases of human life, etc.

Serious attention will be paid to the study of the national history and culture of Belarus, to the problems of development of Belarussian state organisation.

In our view, the implementation of priority directions should contribute to closer scientific and engineering ties with the countries of the CIS and with the Baltic States on the basis of both multilateral and bilateral international treaties and agreements, which today are in the preparation stage by the respective Republican bodies.

In this respect, Belarus already has some experience. An agreement has been concluded with the Ukraine on scientific and engineering co-operation, a similar agreement with Russia has been prepared for conclusion, initial negotiations are being held with representatives of other countries of the former USSR. The works which still need to be concluded are tri-lateral and multilateral agreements on the preservation and development of a united scientific and engineering space. The Republic of Belarus is a participant in the Convent of the countries of the Black Sea region on co-operation in the fields of culture, education and information (Belarus, Russia, Armenia, Georgia, Moldova, the Ukraine, Azerbaijan, etc.).

Unfortunately, scientific and engineering co-operation with the Baltic states is poorly developed. Taking into account the positive experience of the past, we would like to restore the lost connections with the countries of the Baltic region and with the Lithuanian Republic in particular. The practice of co-operation between the Academies of Sciences of our states in the past creatively enriched the science and culture of the two friendly nations.

The problem of conversion, that is, the reorientation of liberated production powers, scientific and engineering potentials and labour resources of defence enterprises and institutions from military to civil needs, is still among the most urgent tasks of the economy of the Republic. The main reasons for the acceleration of conversion processes in the Republic are connected with the exclusively high militarisation level of the Belarussian region of the former USSR, as well as with a rather powerful and unique - concerning certain directions - scientific and industrial defence potential and its excessive orientation towards the defence purposes of the former Soviet Union.

The enterprises of former defence branches of the industry are in the most difficult state. 28 of the largest industrial enterprises and 15 scientific and research institutes and design bureaux, which manufactured about 90% of all military products produced in Belarus, were involved in the design and production of military machinery and armaments. In 1990, the defence products comprised 64% of their total production volume in radio-technical, 56% in electronic, 60% in aviation and 55% in the device-building branches of industry. Over 80% of the products of the Republic was exported. The priority development was given to investigations, design and manufacture of special computer means, mobile automatic control systems, optical devices, communication systems and means, aviation equipment, telemetric apparatus, transport bases for mobile weapons, radiolocation devices, electronic products and other components of armaments and military technique.

Thanks to practically unlimited financing, participation in implementing various scientific and engineering and production programmes, the enterprises and institutions of the Republic could reach world levels in several of the most important directions of science and engineering.

Crucial changes in the international situation which took place in the second half of the 80s resulted in the softening of military confrontation, in adoption of defence-oriented military doctrines by several countries and, finally, in conversion of military industries. The conversion of the defence industry of Belarus, as the subject of the former USSR, began in 1989, and concerned a limited number of enterprises at its initial stage. However, at the end of 1991 and beginning of 1992, as a result of the collapse of the USSR and the established system of material and technical supply of the Military Forces, it acquired a landslide character, uncontrolled and unpredictable in its nature. The optimal conversion rate, when the process remains under control and does not grow into dissociation, is 5-7% per year, while in Belarus the production drop of military destination goods was 4.5-5 fold during two years, reaching a 5-6 times decline ratio in enterprises of radio technical and electronic industry branches. As a result of a landslide conversion, a blow was inflicted to the whole economy, since the enterprises of the defence branches are its heart. The sharp drop in main production undermined the economic, scientific and engineering abilities of the defence enterprises, including those producing competitive civil products. The scientific, engineering and industrial potentials of the enterprise of military branches remained unwanted. The thoroughly established schools and directions began to disintegrate. A very acute problem arose to reorient the liberated scientific, engineering and production potentials of the defence industry towards civil purposes.

With the aim of stabilising the situation, preserving the science-capacious productions and ensuring production of new technically complicated products, conversion programmes for former defence enterprises were designed and approved for implementation in 1992-1993. The programmes were developed by the enterprises themselves and comprised definite directions and projects with the account of technological, scientific and engineering, financial and resource possibilities of enterprises, investigation of markets, solution of the problems for substituting imports and improving the competitiveness of new products on the world market.

The implementation of conversion programmes and their controllability contributed to the reconstruction of science-capacious production into the production of civil products, to the preservation of scientific potentials and qualified employees from the enterprises of the defence branches of industry. In this case the problems of conversion of defence enterprises were studied in close correspondence with the problems of structural reconstruction of the industry of the Republic in general.

With the account of the scientific and production basis available, the existing potentials of the applied sciences, the priority conversion directions should be: modern electronic elements for civil products, computer devices and automated systems, telecommunication and communication means, consumer and industrial radio-electronic and optical products, medical apparatus, equipment for the processing branches of the agrarian complex, technical complex consumer products.

The measures of state support directed in 1993-1994 to the enterprises under conversion greatly assisted in normalising the situation in the former defence branch. By means of respective decisions of the government, a mechanism was established to provide taxation, crediting and other privileges to priority branches and convertible enterprises, which are implementing strategically important investment and scientific and engineering projects.

The success of conversion programmes depends mainly upon the concise and continuous implementation of both home and foreign conversion policies. The realisation of the home conversion policy relies mostly on the enterprises themselves: efficiency of their operation, correct choice of conversion directions, adaptation to modern market relations. An important role in this process is played by state support of conversion enterprises, as an element of economic policy of the state in the field of conversion, as well as measures taken by it to overcome the crisis trends and stabilisation of the economy of the Republic in general.

Still of extreme importance are the foreign aspects of conversion policy, including:

1. the provision of foreign assistance in the implementation of joint projects, including joint investment projects;
2. scientific, engineering and production co-operation with the CIS countries, first of all, with the Russian Federation.

In October 1992, in Washington, a frame agreement was signed between the Republic of Belarus and the USA concerning the liquidation of implications of emergency situations and the prevention of the dissemination of mass destruction weapons, which laid the foundations for co-operation within the Nunne-Lugar Law. For achieving the aims envisaged by the frame Agreement 6, other pilot agreements were concluded, among them the Agreement on the transition of military technologies and potentials into civil activities, providing Belarus with free assistance, including material and technical means, training and services.

Conversion projects have been developed within the Nunne-Lugar Program. The implementation of these projects continues in accordance with the contracts concluded and obligations taken.

In order to provide further development of co-operation within the Nunne-Lugar Program and with the account of the experience accumulated as a result of the joint projects "Integral-Crass", "BelOMO-Belocorp", "MPO BT-FSG" by the enterprises of defence industry branches in co-operation with the Ministry of Industry of the Republic of Belarus, in 1994, new applications were examined for the projects to be included into the programme of 1995, as well as perspective applications developed for new conversion projects.

Co-operation exists between the industrial enterprises and the Commission of European Communities in implementing conversion projects. In 1994, works began on the project of a national programme for designing and launching the production of modern competitive telecommunication and computer means TACIS-93.

In a general positive evaluation of the fact of providing foreign assistance to enterprises of the Republic of Belarus in implementing conversion programmes, it is necessary to note that the efficiency of co-operation in international assistance programmes is extremely low so far. The projects submitted are examined very slowly, local minor tasks are solved without attracting substantial investments.

Rather successful is the co-operation of defence enterprises of the Republic of Belarus with the defence enterprises of Russia.

The main direction of the works in 1995 -- to finish conversion projects concerning the production of new competitive and science-capacious products, implemented within the frame of structural reconstruction of the industry of the Republic, taking into account the necessity to provide a mandatory level of defence capacity of the Military Forces of the Republic of Belarus and satisfaction of the needs of the Russian Federation and the CIS countries in shipments of certain elements, used in the manufacturing of armaments and military machinery.

It must also be stressed that the measures taken by the state and the experience accumulated during the last two years of implementing conversion programmes and structural reconstruction of industry allowed the stabilisation of the industrial situation. The first stage of conversion is mainly over, that is the stage of transition to controlled, managed conversion. The negative implications of the landslide conversion of 1991-1992 have been softened, we have managed to start structural transformations in industry and to determine close (tactical) and strategic conversion targets. The current situation in the defence industry indicates that a further drop in the production of military products, when most enterprises lack alternative designs of science-capacious civil products which are not below military products in their scientific and engineering level, may result in irrevocable loss of unique scientific, engineering and technology potentials in the nearest future. That is why the closest tactical conversion task for 1995 may be stated as follows: the dominant factor should be stressed in the conversion policy from the complete change of the whole nomenclature of the military products produced for the civil ones, into the parallel development of both civil and military production, i.e. into diversification. In this case, in civil production as well, preference should be given to science-capacious, competitive products, in which scientific and engineering levels are comparable to the military levels. Then all resources, equipment, technologies and highly qualified personnel, operating the military production will be used to the utmost.

Experience in conversion shows that:

- conversion comprises not only the problems of industrial potentials but also of the scientific and research activities and the whole of the economic infrastructure in general;
- conversion not only decreases the threat of war but also promotes social and economic progress;
- conversion is country-specific and depends both upon the interior conditions of the Republic and the external situation in Europe and the world in general;
- conversion is accompanied by the structural reconstruction of the national economy as a whole.

THE TRANSFER OF INTELLECTUAL POTENTIAL: POSITIVE AND NEGATIVE EXTERNALITIES

LEONID F. ZAIKO

Social and economic transformation in post-socialist countries is closely tied to changes in the economic branch and sectoral development, including changes of investment flows, subsidies, state and private financing. One of the most difficult problems these countries faced consisted of supporting the level of development of science and higher education and, accordingly, the choice of a national strategy for the intellectual sector development of the society. By 1995, the Republic of Belarus was at its initial stage of developing the system of priorities of 21st century-oriented intellectual potential. It should be acknowledged that one of the first tasks conferred to the President of the Academy of Science by the President of the Republic was to make a list of the most talented people in science who, unfortunately, were unknown to the public.

As in other post-soviet countries, economic recession in Belarus had a negative impact on the development of science and education. Along with fundamental changes in the system of main economic and social institutes, their content priorities of the not so distant past are being revalued. The main factor of the change in public consciousness is the economic situation. The last five year trends show stagnation in the main sectors and branches of industry and deterioration of the main macroeconomic aggregates (see fig. 1).

Statistical data illustrates that the economic recession is deepening. Its content is not a "classical economic crisis" but evidence of a deep transformational mutation of the entire system. The current state of society is characterized by various rates of change in legal, economic and political institutions and the slow, contradictory formation of new economic and social qualities. The external side of these processes is quite typical - the deterioration of practically all macroeconomic indices. The negative economic situation influences the traditionally important indices of net export, the exchange rate (up to 01.01.1995), aggregate supply. The drop in the index of investments bears a negative impact on scientific research and experimental works, creating practically "zero demand" for know-how, patents and licenses on the domestic market.

Economic restrictions were apparent in the areas of mass science and education. The division of USSR property, the declaration of sovereignty of the former Soviet Republics, left not only dozens of scientists unemployed but deemed unnecessary dozens of scientific centres and institutes. The public and some of its

K. Prunskienė and E. Altvater (eds.), Transformation, Co-operation, and Conversion, 69-77.

representatives paid close attention to this "ownerless property". On the one hand, the scientific sphere is not privatized not only in Belarus but in other independent states as well. On the other hand, the state cannot sustain the necessary level of financing of science (there are more than 40 scientific institutes in the Academy of Sciences alone) and higher education system. Practically all fundamental research works requiring high expenditures on equipment and instruments were left without means to ensure a proper level of development in these spheres of knowledge.

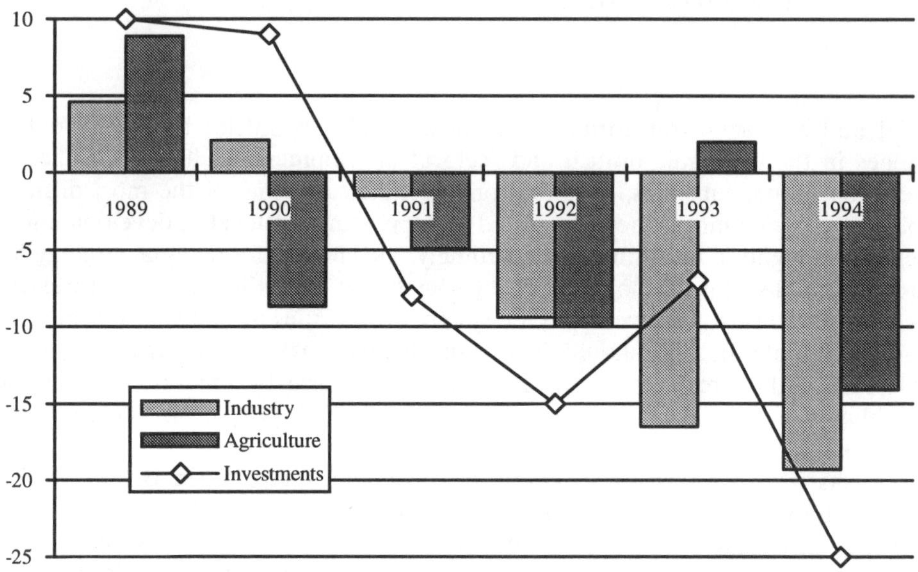

Fig. 1. Macroeconomic indices of the Republic of Belarus (chain indices).

Under these circumstances, the further development of science is becoming problematic with the following situation:

1. migration (including emigration) of scientists,
2. a sharp decrease in the number of employees in science and the scientific infrastructure, causing the preservation of a limited number of scientific schools and directions,
3. the transfer of scientific personnel to new spheres of activity connected with the development of market infrastructure.

We currently face all of these processes. Of course they have their positive and negative aspects and are characterized by positive and negative externalities. As far as emigration is concerned, 480 scientists from the Academy of Science alone left the country during 1994 (data provided by O. Borodinsky). The peculiarity of this situation is that these scientists are still formally employed by different institutes. Thus, out of the total number of scientific workers, 4747, 10% of scientists left the scientific associations of the country. This index is very important, since the most promising and qualified scientists left the Republic. In prospect, this transfer will decrease alongside the reduction of leading intellectual potential. However, the negative externalities are quite vivid: a loss of 3 - 5 leaders in any branch of knowledge throws the development of a whole perspective direction in science back 15 - 20 years.

As with the "external" transfer, our younger generation of scientists leaves this sphere in the direction of "interior" transfer. Only 13.53 percent of postgraduates took jobs in the Academy of Science. 1995 was the first year that the Academy did not have a single Doctor of Sciences under 30. The material position of scientists is unsatisfactory with a steady declining tendency. On February 1, 1995, the salary of the Director of an Institute equalled 12 minimal monthly salaries, and from March 1 (while an official increase of wages was declared), it decreased to 7.64 times the legally defined minimum. As for the reduction of the scientific sphere, this process is without doubt of great importance to Belarus. Its exterior features are apparent from an analysis of statistical data for the number of employees in the scientific sector. (see fig. 2)

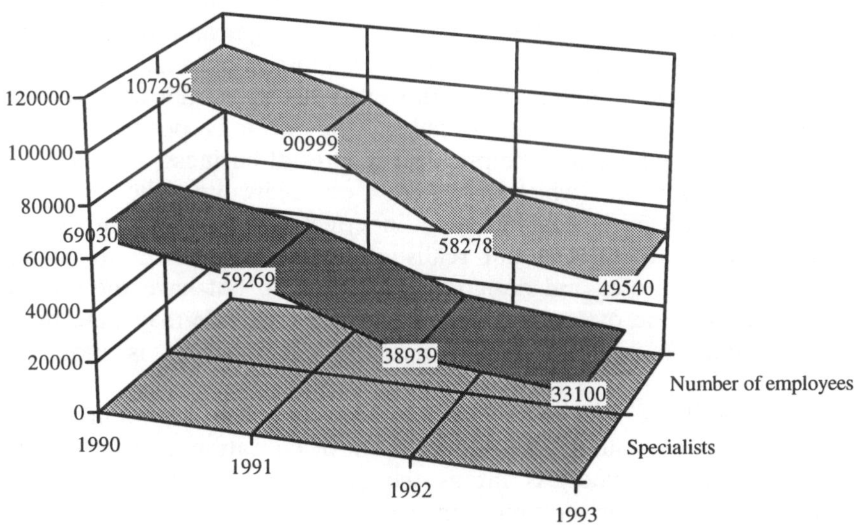

Fig. 2. Dynamics of number of employees in the scientific sector

As the chart indicates, the reduction in given categories of scientific employees is double, a process which still continues. For example, the number of employees in the Academy of Science decreased by 14.53 percent. Quantitative indices conceal qualitative sides of the process. The characteristic feature of this is that the majority of those scientists who leave is quite young and promising. These scientists choose other spheres of activity: business (small and medium as a rule), trade, banking, finance, export-import operations. This process is negatively estimated by society (first of all by scientific circles), though this approach is one-sided. This kind of migration does cause positive externalities in the development of national gross intellect in new spheres of practical activity. There was also a time in the USA when relative "overproduction" of specialists in the humanitarian sphere caused their migration to "production", management, business, giving new impulses to the development of labor management systems and the creation of more dynamic and efficient organizational forms of production and management. This had a major impact on all spheres of society for decades.

As for a majority of post-socialist countries, the interior migration of scientific personnel may cause sharp changes in the general intellectual level of leaders in business, finance and applied technologies. The content and speed of this process in the history of transitional societies are unique. Moreover, it is becoming a powerful catalyst of economic renewal. Intellect and knowledge are transferred into practical spheres, reinforcing potential resources in the development of new sectors of business activity. This new market sector, however, is also formed from representatives of semi-criminal and criminal structures. And it also influences inter-relations in the business sphere. What seems stronger and more promising: intellect multiplied by decency and ethics or an aggressive criminal world with its behaviour similar to that of animals? The answer to this question is not in favour of this or that situation. As for our Republic, the score is positive for those businessmen who left science for the practical spheres of business. They are called for either to start production (which is a paradox, especially under the conditions of overproduction which have recently been in effect) or to return to scientific laboratories and institutes. It is not realistic for them to return to collapsed scientific research institutes and design offices. They do not have the money to finance the same volume of research works as in the past. We may state that the interior migration of intellectual personnel to new spheres is most likely irreversible.

Economic depression has a manifold impact on the development of the basis of the intellectual foundation — the system of education. A few tendencies and trends can be observed. As far as the general situation of education is concerned, the first changes took place in the form of a decline in the number of students. From 1985-86, the total number of students was 3,913,000 (in all forms of education). By 1993-94, this dropped to 2,412,000. A sharp decrease in the number of students in the system of personnel training at enterprises (concealed unemployment makes this situation natural) and an increase of qualifications took place. Some Institutes for Teachers Training ceased to exist or were transformed into private business schools. The only sphere in which the number of students increased was in the secondary schools. This is due to the demographic situation and obligatory secondary education.

The dynamics of changes in the number of students in various spheres of education show the following tendencies: (see fig. 3)

A decrease of the number of students at HEIs (Higher Educational Institutions) became apparent by the 1993-94 school year, though this decline was not substantial. Simultaneously, the financial possibilities for the educational sphere are vividly decreasing. The same situation is apparent with the material well-being of HEI teachers. For example, in March 1995, the average salary in the Republic was 700,000 rbl., while that of a professor was 590,000 rbl. I should add that it was the lowest material status of scientific workers and professors during the entire period of socialist and post-socialist development of the Republic. We cannot state that this automatically makes these teachers opponents to reforms (or the imitation of reforms), but it is quite natural that the attitude of a majority of professors, university teachers and scientific workers of the system of the Academy of Science is negative toward "restructuring processes".

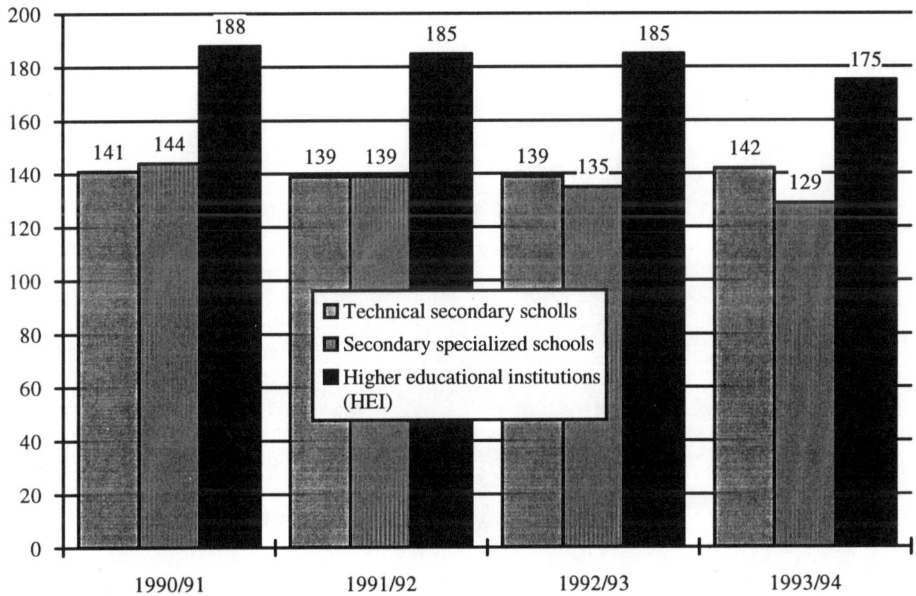

Fig. 3. The Number of Students per Type of Education (thous. people): Technical Secondary Schools, Secondary Specialized Schools, Higher Educational Institutions (HEI).

Some interesting changes are noted in higher education. Firstly, there is a slight decrease in the number of students but there are more graduates, i.e. specialists with higher education. (fig. 4).

Secondly, the growth in the number of specialists per 10,000 people of the Republic was observed. This trend shows inertia in the development of the educational sector. There is no economic necessity for such a number of HEI graduates. The unemployment figures of those with a higher education degree are no state secret, but different social groups still follow the tradition of sending their children to universities and institutes. Parents are likely to motivate their behavior by the very dubious hope that a higher education will be a "permit" to new economic and political elites.

At first sight, the general tendency might be estimated as positive: There is a growth in the number of specialists with higher education in spite of economic recession and considerable changes in the social stratification of society. The value of education remains a factor of public consciousness, though new priorities (money, real estate) modify the youth's values to a great extent. This may be considered an inertial element and proof of keeping to the fundamental characteristics of public values even during the transitional period. Higher education in any society is a definition of social identification, a "sign" of belonging to a certain political or economic elite. However, in the world as a whole, only a few of the most expensive and representative universities pave the way to high society. As for Belarus and Minsk, we do not have such institutions yet, although we can observe first shy and not very successful attempts in their establishment.

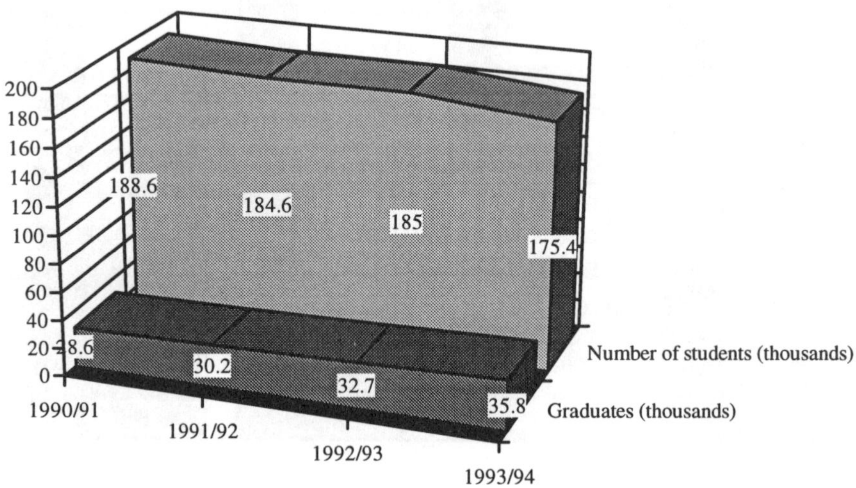

Fig. 4. Total Number of Students and Graduates of Higher Educational Institutions (HEI)

Interior transfer also took place among the scientists of the system of education. The most dynamic young economists and lawyers quickly found new jobs in fields of practical activity. Representatives of other professions immediately reacted to the situation: many mathematicians turned into economists and financial managers, physicists and radio mechanics became real estate and liquidity market specialists. Thus many people began to learn new professions to intuitively and semiprofessionally fill vacant positions in the market system. This sort of effect is both positive and negative. First, people found well-paid jobs. On the other hand, the absence of professionalism or semi-professionalism in the economy creates additional chaos and misinterpretations. This is vivid at all levels of the society. Semi-professional politicians and economists will surprise us for a long time by their discoveries in the history of the Republic, legislation and technologies of democracy, economic theory and practice.

HEI staff is one of the powerful intellectual spheres which does not give up, enduring all blows of fate, and its ranks are as firm as they used to be even after the transfer of young specialists to other spheres of activities. The number of university teachers in Belarus has not decreased recently (see fig. 5), though qualitative changes have taken place in the activities of this intellectual elite.

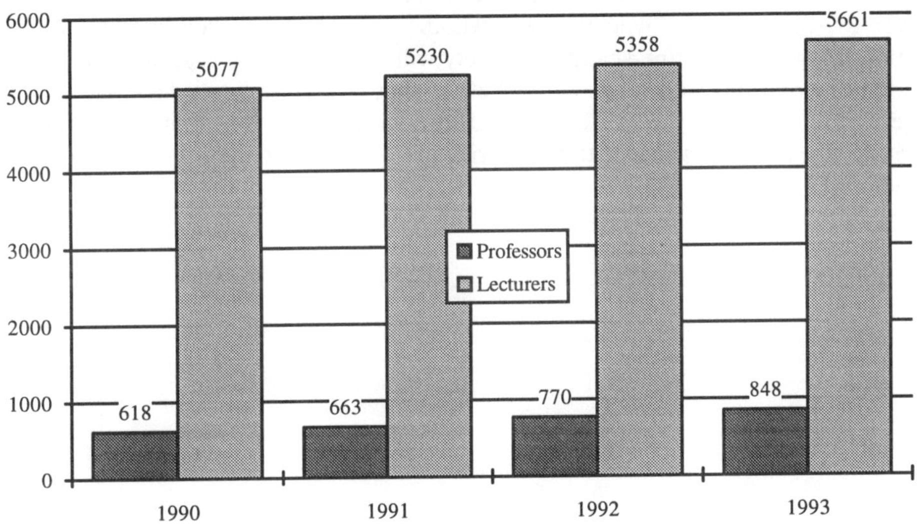

Fig. 5. The Dynamics of Teaching Staff

The number of professors and lecturers in HEIs lies between 5,600 and 6,500. It is significant that this number is steadily increasing despite the deteriorating material well-being of scientists. Unlike the scientific research sector, the number of HEI teachers has not decreased, and they still occupy the same niche in the general employment structure of this sector. There is a probability that this situation will be stabilized or "conserved". These facts are often linked to external visible faucets of the problem. Interior changes are more important. These are related to the transfer of young prospective scientists and teachers from the educational sector. It is quite possible that professorship will be represented only by those scientists who received their degrees in the 60s- 80s. We should keep in mind that Scientific Councils and Departments are overwhelmed with specialists of our "historical era".

The irresistible desire to create new scientific societies and academies and to give each other honorable degrees and titles is the last stage of hyper-compensation for the representatives of the science of the past. At the same time, the most talented are leaving for the USA, Germany, France and other countries, where such degrees and titles (in our understanding) do not exist.

Apart from the HEI sector, a large number of scientists is employed in the academic sphere. Its personnel structure is depicted in fig. 6.

What is peculiar about this scientific association? Scientific elite, as in any other sphere, amounts to 3 per cent. Its existence is not problematic. It consists of old retired or old generation scientists who are about to retire. Doctor of Sciences (there are only 444 of them) are also not likely to change jobs or spheres of activity. Otherwise they would have done what their famous and authorities-caressed former President mathematician did. Instead of leaving for the world of capital, scientific hierarchies let capital in. To be more precise they let the banks in. The answer to the question "who will win whom" might have a well-known Lenin's solution. However genuine scientific society will survive in any case whether there are degrees or titles and money or whether they are completely lacking. In this sense we are unique in the world history. But nobody knows how this 97 per cent of scientific workers will survive: their salaries are too low for them to afford decent clothes and food.

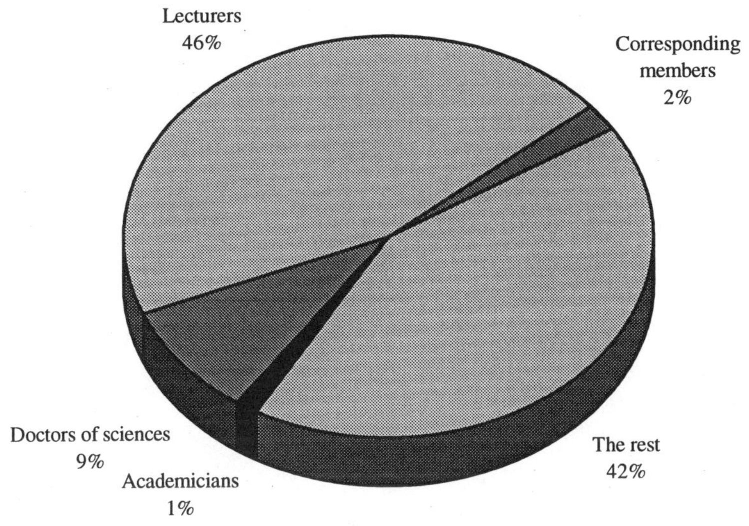

Fig. 6. Structure of Employees of the Academy of Science of the Republic of Belarus

In general, the Republic of Belarus is on the edge of an intuitive choice of a national strategy to develop the intellectual potential of the nation. Influence of exterior and interior factors of development of intellectual resources and their implementation is under focus of political discussions. At present, this problem is considerably influenced by the revival of stereotypes of "socialist romanticism". A general tendency of change in the strategic course of the new leadership of the Republic also proves this. Official policy referring to scientists and teachers repeats the patterns of the past. Due to this fact, it is difficult to conceive the breakthrough of the "vicious circle" of development of science and education in the transitional society of the Republic.

MASTERING RUSSIAN TECHNOLOGIES AND THEIR PRESENTATION TO THE WESTERN MARKET: THE POSSIBLE ROLE OF THE BALTIC COUNTRIES

V.D. NOZDRACHEV

The analysis conducted by UN experts in estimating the activities of the International Monetary Fund (IMF) and the International Bank of Reconstruction and Development (IBRD/World Bank) has not only proved the low efficiency of financial support to developing countries but also a market tendency in preserving their economic backwardness.

Some of the measures proposed by these organisations are rather rational and thus are supported by economists of developing countries. At the same time, however, the IMF and the IBRD make some serious mistakes (though one cannot refute their purposeful intentions and well-considered actions). One such mistake is the application of a standard scheme which does not take into account all of the peculiarities and conditions of an assisted country. The principal mistake of the IMF is that its programmes are aimed at demand suppression instead of economic development with production expansion leading to income growth. "Shock therapy," resulting in impoverishment and higher social tension is no less dangerous. Its authors appear to know nothing of the ideas of ethnography written by Gumilev, concerning the so-called *passionis* (Lat.; the English equivalent of this term is *drive*). The lack of attention to this phenomenon in regions where large scale experiments by the IMF are conducted is fraught with social disturbances, the consequences of which may be fatal for the USA and other leading Western countries behind this funding.

The refusal of grants and subsidies to the most important branches of industry and agriculture may be disastrous for national economies as well. The experience of Hungary and Bulgaria (the agroindustrial complex) demonstratively exemplifies this.

Where Russia is concerned, in the opinion of independent experts of third countries, the IMF and the IBRD programmes are not aimed at the development of its economy but at the prevention of a collapse regarding such external economic obligations as payments of foreign export invoices, profit taxes from investment, the repayment of credits, etc., giving way to a further worsening of the situation. Moreover, the IMF recommendations on import liberalisation for the sake of export growth will result in the export of irreplaceable resources exchanged for Western consumer goods. This may be regarded as indirect financing of the

K. Prunskienė and E. Altvater (eds.), Transformation, Co-operation, and Conversion, 79-83.

Western economy at the expense of the accumulation of currency means (belonging to Russia in particular) by the accounts of Western banks.

The strengthening of integration of the CIS, the countries of the post-Soviet period and Eastern Europe contradicts the interests of the IMF, the USA and large European countries. The restoration of closed production cycles of the past will reduce the need for imports, lower the prices of goods and support the economies of many countries. Of course, this does not correspond to the interests of national and trans-national corporations supporting the IMF.

Some UN experts consider it necessary to pay attention to the dynamic development of the Eastern-Asian region where the IMF recommendations have never been applied. Despite the fact that the growth dynamics in every country have been similar, one should take into account the tremendous differences in their economic and political starting conditions (for example China, Japan, Indonesia, etc.). Some identical trends in all programmes in South-East Asia are evident:

- special emphasis on export development;
- the gradual introduction of changes;
- the preservation of state regulating functions;
- moderate protectionism;
- the expansion of special programmes or their support;
- attention given to the specificity of a country.

Analogous tendencies are noticed in Latin America as well, where such countries as Chile, Argentina and Mexico have abandoned the policy of "market extremism" of the IMF and have developed their own programmes. It is interesting to note that a common feature of all countries carrying out successful reforms is the refusal of the privatisation of the main export branches (copper in Chile, oil in Mexico).

Models imposed by the IMF may bring positive results only in small countries, where the presence of raw materials fit for export is concerned (Costa Rica, Ghana, Estonia). An inflow of foreign capital into production should be stimulated only if there is a possibility for the national industry to have access to the latest technologies.

There is a contradiction. It is not possible for large Western companies to introduce these technologies, as they lose the mechanism of "undressing" various countries worked out by the IMF for their benefit. At the same time many technologies undergo the process of improvement and become more "compact" from a spatial point of view and less energy- and resource-consuming. This causes the formation on a second industrial wave. Large, and as a rule conservative companies, begin to lose their specialists and some links in their technological chains or are unable to eliminate obsolete technologies in time and without considerable losses. Specialists transfer to medium-sized but developing companies which possess break-through technologies. Some of these companies have no scientific potential of their own. Thus, they are ready to buy specialists and developments or to invest money into new branch establishments through which the results of investigation and technologies necessary for them can be acquired. Pools of such companies are quite competitive with those of larger companies. Besides, the latter must actively defend themselves from any leakage of information and specialists. At the same time the military-industrial complex of the former USSR, or to be more exact, its remnants in the countries of the CIS, possesses unique developments and technologies. Let us look at some of the problems in connection with them.

Today Western countries attempt to purchase as many specialists and developments as possible. Taking into account the absence of experience, juridical protection and the miserable existence of research institutes and design offices, this is done for kopecks. However, there are certain reefs in this process. The West possesses information on some leading scientists and institutes. For example, the rating of the Saint Petersburg Physical-Technical Institute is rather high, and many of its specialists work successfully in various Western countries. But at present this Institute has become a starting-point for many quite ordinary scientists, whose only goal is to leave the countries within six months to a year of their appointment. The specificity of academic selection, the approval of theses written with the purpose of receiving their scientific degree, the nomination of this or that post in the former USSR, resulted in the appearance of a considerable number of high ranking persons, who were not the best representatives from an intellectual point of view. Thus, a greater part of the true bearers of know-how and technologies was not mentioned in scientific reports and remained unknown. This was also typical of branch institutes, which worked on the applied aspects of developments. One may conclude that the sale of ideas and developments carried out by scientists and officials of high standing is not a guarantee of their realisation in the West, even if this is accompanied by the transfer of technical documents.

This tendency has given rise to one more phenomenon, characterised by the leakage of ideas and scientists to South-East Asia, the Arabian countries and foreign organisations which strive for the spreading of their ideas by force. It may seem strange, but the majority of true holders of know-how prefer this. The most unpleasant aspect of this process is that many of them are guided by ideological causes, witnessing the impoverishment of their compatriots and the enrichment of inspirers of this policy abroad.

Owing to the above-mentioned tendencies, preconditions appear for the realisation of a situation, which allow various parties mutual profits. Capital possessed by mainly medium-level companies can be invested into high technology branches, complex technologies and unique developments with excellent consumer properties. Moreover, Russia has a considerable hidden potential. Investment into projects which are not connected with raw materials (oil, gas, etc.), is extremely risky because of the policy of the IMF, the current state of the political situation and the economy in Russia.

There is also a third factor which may help solve the discusssed problems. This factor is the Baltic countries and their geopolitical position. The sharp weakening in the role of these countries in the policy of the West, aimed at the destruction of the communist system after the disintegration of the USSR, has led to quite a natural reduction in financial assistance. Today only Lithuania can hope for an increase due to its geopolitical position, the regional interest of Germany and a number of contradictions between some developed countries. This can only be achieved if the political priorities are properly made (although experts have so far failed to see such priorities in Lithuanian foreign policy). Nevertheless, it is just this very geopolitical position, which is responsible for a higher level of stability in the Baltic countries reducing the risk of investments into high technology branches, which is of special concern to Lithuania.

A higher cultural level of production in comparison with the countries formed during the post-Soviet period as well as a level of general education of the population which is much higher than in some developed Western countries, should be regarded as very positive factors.

Thus, the most expedient scheme for a solution to the above-mentioned problems is the participation of medium-level companies in the process of investment into the Baltic countries, with the purpose of mastering conversion technologies and developments, which come from the countries of the CIS. This will also allow the expectation of considerable investments from Russia. This is an extremely profitable process for the Baltic countries (commitment of the population in production, demanding a high scientific potential, acquiring a "technological" face, the weakening of the IMF and its influence, etc.), the countries of the CIS in general and Russia in particular (a constant and considerable inflow of money from technology export instead of occasional payments which usually end up in somebody's pocket, the stabilisation of prospectives and extremely complex trends, the "binding" of the most valuable part of the intellectual potential), Western countries and their medium-level business people (more stable regions for investments, an access to the latest technologies, control over potentially dangerous technologies, a smaller leakage into countries with extremist orientations, etc.).

The USA and some other countries will try to oppose this process with the help of both political and economic measures. This is quite natural, since the development of break-through technologies in compliance with the above-mentioned scheme will weaken the influence of the USA and some of its large companies in Europe.

Resource-dependent technologies which, for instance, depend on receiving new material for radioelectronic components on the basis of nitrogen-free diamonds, may serve as an example of such key technologies. Taking into consideration unique electrophysical properties of natural nitrogen-free diamonds and their considerably lower cost in comparison with artificial analogues, one may say that countries and companies controlling these resources and technologies will be able to revolutionise radioelectronics by the beginning of the next century. It is possible to draw a parallel between this revolution and the replacement of valve technology by semiconductors. As a result, they will control prices in the radioelectronic products market and modern arms.

The existing economic situation provides a unique chance to the Baltic countries, countries of the CIS and Western Europe, to make a considerable step forward. Let us hope that politicians possess enough wisdom to give up their political ambitions in favour of the prosperity of their peoples. Otherwise we will face some serious problems from the Middle East and South East Asia, which will do everything for the quick acquisition and application of break-through technologies, in contrast to the self-satisfied and satiated West.

THE SCIENTIFIC-TECHNICAL POTENTIAL OF THE KALININGRAD REGION AND THE CONVERSION OF ITS INDUSTRY

B. SAUSKAN AND C. OVERKO

The Kaliningrad region, as an enclave territory of Russia, was established in December 1991, after the fall of the Soviet Union. Since this time, completely new social and economic problems have arisen in various sectors, among which the most important appears to be the guarantee of employment and economic vitality under radically new geopolitical and socio-economic situations of the region.

Reasons for the appearance of these problems lie in the economic crisis which inflicted our region on a larger scale than all of Russia. First of all, the transport ties in the region sharply worsened, compared to other regions in Russian and the CIS, from which up to 90% of raw materials are obtained, as well as manufactured goods. Secondly, there was a severance of economic ties with the territories of the newly-independent republics and, thirdly, - the common problems of the transition period to a market economy.

The industrial structure in the region is still comprised of three main branches: fishery (21.7% according to production volume at current prices), wood processing and pulp and paper (15.3%), machine production and metal processing (11.5%).

According to the federal development programme of the Kaliningrad region from 1995-2000, priority development of the following complexes is foreseen: transportation, fisheries, wood processing, agro-industry, machine production, health-energy, construction, and the social-recreational complex. The following are also considered important tasks: ecology, development and implementation of the social sphere, development of external economic ties and the formation of local free economic, financial and off-shore zones.

The crux for development of the above-mentioned sectors will be the scientific-technical and production potential in our region.

Historically, we have sufficient development of scientific-technical potential for such a small territory as the Kaliningrad region, comprised of scientific-research institutes, a construction bureau, scientific-industrial establishments, the fishing industry and geological production research institutes, scientific-research laboratories and experience in experimental production.

The structure of scientific-technical potential in our region consists of, on the one hand, the industrial and economic structure, and on the other hand, its specific coastal situation.

K. Prunskienė and E. Altvater (eds.), Transformation, Co-operation, and Conversion, 85-89.

Over half of the scientific-technical potential in the region was employed toward solving problems in the naval sector for the Western USSR: fishery, water transportation, ship construction, sea ports exploitation, sea product processing and all coastal infrastructure.

Apart from the naval complex, scientific research and processing in our region was implemented in the following directions:

- implementation of technology of agricultural production;
- the ecological situation of the hydro-sphere, atmosphere, lithosphere and biosphere in the Kaliningrad region and problems of environmental protection;
- fundamental research in the region of oceanology, physics, mathematics, history, philosophy and law, literature and philology, pedagogy and education;
- research and implementation in the machine production sector, ship construction, parts production, materials research, the automatisation of manufacturing and energy;
- theory and practice in market economics and business;
- theory and practise in administration and management.

The potential of cadres in science and technology in our region is sufficiently high: over 800 candidates in sciences, over 100 doctors of science, 10 academics in various academies, which work in more than 30 scientific-technical educational, production and administrative organisations in the Kaliningrad region.

However, the disintegration of the Russian economy during the 1992-1994 transitional period was also fully reflected in the scientific-technical potential in our region: scientific research decreased almost by 50%, the number of employees in this area declined by over 60%. The privatisation of certain scientific organisations led to their partial or total re-formation. Due to economic reasons, the previously technically equipped scientific-technical ports "AtlantNIRO," "AO IORAN" which employed many workers, noticeably decreased their naval research works and began conducting commercial activity or work other than according to their sphere of activity.

For the preservation and development of scientific-technical potential, a series of methods of organisational character was accepted: in March 1993, in the administrative region, a department of science and higher education was established. A series of scientific-technical programmes are being formed under this initiative, as well as projects of regional character, regional scientific-technical conferences are being conducted, a scientific expert-consulting council of representatives of the higher education, and scientific research institutes of famous scientists were established.

Stronger collectives of scientists are currently concentrated in higher educational institutes, and also in the Atlantic Scientific-Research Institute department of the fishery and oceanography (AtlantNIRO) and the Atlantic Department of the Institute of the Oceanology Academy of Sciences of Russia. These collectives support scientific contacts with scientists from Poland, Germany, the US, the Ukraine, Lithuania, Estonia, Denmark, Canada, Great Britain and other countries.

The number of scientists coming to Kaliningrad from the Baltic States, Kazakhstan and the Northern Caucuses, which are prepared to work not only in the scientific sector but also in industry, management and business, is increasing.

In 1994, for the first time, a regional scientific-technical programme was established and realised, featuring seven sub-programmes:

1. the scientific, methodical and informational preparation of environmental protection specialists;
2. the supply of scientific-technical information for the development of the Kaliningrad region production complex;
3. economic research of the Russian enclave region;
4. the social, humanitarian development problems of the enclave Russian territory;
5. new technological and technical decisions;
6. the development of the transportation complex in the Kaliningrad region as the most significant transport centre in Russia;
7. regional information on the education system and science in Kaliningrad.

This programme received financing from the Ministry of Science, and its separate sub-programmes have already been implemented.

The prospects for further development and the implementation of scientific-technical potential in the Kaliningrad region, along with the scientific-technical policy of the Russian Federation, are comprised of the following:

- the establishment of regional co-ordinating, innovative and other centres in scientific-technical spheres;
- the establishment of European-level scientific educational complexes;
- the creation of four techno-parks (1 - for machine production and metal processing, 2 - ecologically-safe technologies, 3 - development of the naval economy, 4 - socio-economic development), joined in the Russian international technological centre;
- the establishment of specialised regional investment funds;
- the creation of a non-government scientific-technical sector;
- the establishment of conditions for loans, leasing and sub-leasing, payment for communal services, for organisations in the scientific-technical spheres;
- the implementation of a system which materially encourages scientists and specialists in the scientific-technical sphere;
- the creation and acceptance by the local Duma of a law on scientific-technical policy in the Kaliningrad region.

This is the situation and these are the development prospects of the scientific-technical complex for academies and higher educational institutes in the Kaliningrad region.

Apart from these complexes, the main role in regional economy is played by the factory complex, which combines construction bureaux with production organisations of various forms of ownership ("Stroidormash", "Vagonzavod",

"Svetotechnika", "Sceptres", "Kaliningradgazaftomatika" and others). These enterprises manufacture goods for the construction of ships, machine production, wagons, the electronics sphere, forestry and others. Until 1991, approximately 30% of these firms implemented orders of the USSR military industry.

On March 20, 1992, Russian Federation President, B.N. Yeltsin, signed a law on the conversion of the Russian Federation defence industry. This law outlined the legal foundations for national defence activity and related conditions for enterprise and organisation for the reduction or total elimination of military industrial orders as well as their production capacity conversion, scientific-technical potential and labour resources. On December 23, 1992, according to order Nr. 304 by the Kaliningrad region administration, the regional conversion programme was signed. This programme concerned, first of all, nine converting machine-production enterprises, which, in 1990, comprised 50% of all military industrial production. The main task of this conversion programme was the utilisation of accumulated high-technology of the military industry in the production of high-quality consumer goods. Conversion programmes are connected with the development of the free economic zone in the Kaliningrad territory. Following this, the following main priorities for the possibilities of utilising conversion enterprises, were accepted:

- production, storage and processing of agricultural and sea products;
- the increase of the material-technical base through the provision of communal services;
- the increase in nomenclature and production volume of local goods and especially complex consumer technology;
- the development of sea ports and the transportation system;
- the construction of ships for government purposes;
- the production of construction materials from local raw materials.

From 1993-95, all of these enterprises decreased their production of military goods. Currently, according to a law of the Russian Duma, the local Kaliningrad government has the right to issue credits only for the manufacturing of social goods: perishables, children's apparel, school supplies, medicinal equipment, consumer technology, medicinal services, public transportation and the construction of residential housing. In 1995, the part of military industry orders to converting enterprises in the Kaliningrad region did not exceed 15% of total production.

If we are to speak of specific enterprises, then AO "Kvarts" began producing consumer washing-machines, portable medical equipment for nervous system disorders, gas generators, and vacuum packaging machines.

AO "Jantari" ceased production of military ships and now manufactures passenger and transport ships.

AO "Stroidormash", instead of its earlier-produced civil and military defence technology, now produces only civilian automobiles.

AO "Fakel" - autonomous energy-saving complexes, medical equipment.

AO "Kaliningradbumash" - pulp industry equipment.

Besides the conversion programmes in the region, suggestions are prepared for the diversification of production, mainly for the enterprises of machine production and metals processing, related with new conditions for the transition to a market economy.

The essence of these plans is the transition from the production of large metal construction to low metal requiring metal production as well as consumer goods production.

We may single out the following diversification directions for production:

- complete growth in electrified transportation;
- the production of progressive technologies, including electro-reactive motors and space shuttle motors of the future;
- agro-industrial machine and equipment production, for use by commercial and individual farms;
- passenger vessel construction and naval containers;
- an increase in the production of goods for the fishing industry;
- production of medical and communications equipment;
- production of high-consumption goods and health/athletic inventories.

Also, the Russian (mainly the Kaliningrad region) conversion programmes are followed by a number of difficulties, such as the re-training and employment of soldiers, the solving of their social problems. Certain issues in the Kaliningrad region are being settled, however, due to the economic difficulties of the transitional period, finances are insufficient for the more rapid finding of solutions.

Currently, there are practically no orders from the Ministry of National Defence. Military academies are being transferred from the Kaliningrad region to other regions. The Baltic military fleet has been decreased by almost double. Missiles have been completely removed from the region, and the land divisions have also been noticeably decreased. All of these means were taken in order to lessen tensions in the Baltic region and to develop peaceful co-operation among these countries.

The specific geopolitical situation of our region, on the one hand, creates additional transportation, taxation and customs difficulties for the economy. On the other hand, it creates possibilities for closer co-operation with other countries, especially the Baltic region, which will especially assist in the development programme of the local free economic zone, off-shore and perhaps financial zones.

Along with this, the main sector of economic development should become scientific-technical and industrial potential, without which it is impossible to produce economic and competitive goods on the external and internal markets.

We hope for closer and mutually beneficial economic and scientific-technical co-operation with the Baltic states and are prepared to participate in various regional and inter-regional programmes with the goal of Baltic regional development.

THE UTILISATION OF UNIVERSITY POTENTIAL AND CO-OPERATION IN EUROPE

P.J. TEUNISSEN

This contribution will first focus on a general analysis of international educational exchanges in Europe. On this basis, policy suggestions will be made, including suggestions for the Baltic States.

In the past ten years, international educational exchanges have become a large scale activity in Europe. This has resulted from a number of impulses which changed our educational system in rapid sequence.

In the 1980s, the European Community launched a series of exchange programmes in order to create an infrastructure in the spheres of vocational training, education and science for European integration. The most important was the Erasmus programme, which stimulated the establishment of several major networks of European universities. Also, the total number of international Co-operation Agreements of European universities now amounts to about 2600.

After the fall of the Iron Curtain, the European Community initiated the TEMPUS programme for exchanges with Central and Eastern Europe. Exchanges with the Commonwealth of Independent States were promoted under the TACIS programme.

The historic developments in Europe evoked a new wave of exchanges between Europe and other parts of the world, in particular with North America.

At present, the internationalisation of science and education in our region is receiving additional dynamics from the revolution in information and communications.

Building on this, the EU has recently adopted the Leonardo da Vinci programme for the internationalisation of vocational training and the Socrates programme for international co-operation at all levels of education, from primary school up to university. Socrates incorporates Erasmus and other programmes. The EU now disposes of two comprehensive framework programmes for the internationalisation of education. The management of these frameworks will be shared with national liaison offices. In 1997, universities and high schools will have to conclude an Institutional Agreement with the EU in which their exchange programmes will be registered, and they will be held responsible for their implementation. In this way, the Union will receive a real educational policy, which will provide a new momentum to integration.

K. Prunskienė and E. Altvater (eds.), Transformation, Co-operation, and Conversion, 91-94.

Positive as these trends may be, we should at the same time note the opposite trend of the renationalisation of thinking, which dates back to the 1960s and was reinforced by the end of the Cold War. This trend is reflected in the minds of the students, by a fundamental lack of knowledge and ideas about Europe and international affairs. The foreign students we receive are generally interested in the subject of Europe because they know it is important. But they have no vision of the future of Europe. In this context we should also point out the fact that in several countries the teaching of history in primary and secondary education has been drastically reduced. And the universities are under financial pressures to compress their study programmes at the cost of broader intellectual and cultural education.

A factor of major importance is the absence of a European culture. What we do have in Europe are common values, common arts, in some respects a common lifestyle. But the diversity of languages, of legal traditions, the religious divisions in Europe and the strong national and ethnic traditions exclude the existence of a European culture in a meaningful, social sense, for the time being.

European integration is, for the major part, a goal that rests on economic and political or security arguments. It is, as yet, much less an aspiration of the peoples of Europe.

Integration will continue, most of all due to the dynamics of modern society. The essential question is whether Europe will become a region of democracy and justice, governed by strong institutions and united by the solidarity of its peoples; whether Europe will at the same time be open to the outside world, including its neighbours on the Eastern side.

In the past few years, it has become clear that entry into the Union of the associated states of Central Europe will require important changes and long-term common efforts on both sides. In this undertaking, the educational institutions will have to play an essential role. But the present exchanges are not yet sufficient to meet this task.

Some policy suggestions. If feelings of European identity and solidarity are to be fostered, education in modern European history and European integration must take place at all levels of scholarly education. The European Union has become quite active in this field, but it remains dependent on the member states as far as educational practise is concerned.

The associated countries should seek international exchanges at the widest possible range of levels of education. They should preferably not focus on university exchanges alone, but at least on exchanges in the sphere of technical vocational training as well. This sector has become the most dynamic one in our educational system, and it is of vital importance to the modern economies, offering the most opportunities for combining studies with internships in the economic sector.

The associated countries should also seek consultations with the European Union on this matter, on the basis of the European Council declarations of

Copenhagen (June, 1993) and Essen (December, 1994). In these consultations, they should include the question of the mutual recognition of diplomas and other professional certificates. Since the Maastricht Treaty of 1991, the European Commission has become competent in this field and intends to adopt new initiatives.

The universities and high schools from Central and Eastern Europe still lag behind in their international exchanges. It is not clear to what extent they could join existing networks. Networks may become unworkable by expanding too much. Therefore, it may well be that new networks will have to be established.

Exchange contracts of networks should be established or consolidated ultimately in 1997, when they have to be submitted to the EU for inclusion in the framework programmes.

Exchange relations should be considered from the perspective of their long-term contribution to the external relations of countries. For the Baltic States, developing closer relations with the Benelux is relevant. To some extent, the Benelux can indeed function as a model. Benelux is an area of transit communications, consisting of smaller nations which operate actively in Europe and in the world. These smaller nations have learned to carefully develop their mutual relations, which are a mix of affinity and clear diversities in the religious and cultural fields.

Yet, as far as Europe is concerned, the Balts would do well by also seeking exchange relations with the European "big three": Germany, France and Great Britain, since these countries will have the strongest influence on the shaping of the future Europe.

Narrow programmatic training should be avoided. Students studying abroad should be given opportunities for general education. They should have time to write a paper or to participate in projects of the host institution. University students should as a rule stay abroad at least half a year. In practise, this would require an improved co-ordination of study programmes.

Preferably, foreign students should not only be housed in international student houses, but also among the population of the host country or in houses of national students. They should be given the opportunity to mix with students of the host country and to participate in activities of their student associations.

International exchanges must reach a minimum level of intensity and commitment if they are to be effective in a broader sense. The bridging of national barriers requires deeper common experiences of having lived and worker together.

A further aspect concerns the Brussels bureaucracy and the criteria for receiving EC support. The EU tends to support only those exchange programmes in which at least two European Union countries are involved. Before programmes with Central Europe can be started up, partners must be sought in at least one other EU country. This often is too burdensome for members of the university staff who are interested primarily in seeking partners in Central and Eastern Europe. In Groningen, we have been trying to establish contacts with a university in the Ukraine for four years. So far these efforts have been in vain because of defective communications and the complexities of the procedure. The European Union should be more pragmatic here. It should support exchanges on a bilateral basis, under the condition that they expand in a later stage.

The Socrates programme of the European Union clearly focusses on the telecommunications revolution and long-distance education. It gives less priority to co-operation agreements between universities as they are existing now. This probably has to do with scarcity of financial resources as well. A clear warning is necessary here. A telecommunications network among European universities may be very helpful for research and may support teaching activities, but it cannot offer the general solution for higher education. Nothing can replace the direct contact with foreign countries and the local people. And students are crucially dependent on direct contact with the teachers. The modern trend towards standard study packages and standard examinations is very harmful to the university. Universities are in the forefront of scientific education, where the problems and the solutions are not yet clear. And university students are no learning machines. They are young people who still have to find their way in life and are in a critical phase of the development of their personality.

The programme for Jean Monnet Chairs in European law and European studies is highly valuable indeed, but it cannot replace direct international exchanges of students, for the reasons indicated above.

Our universities should undertake more common research projects relating to the future development of the EU, including its extension and its external relations. Such projects are still too scarce. The main think tanks in this field are in Brussels and London, and they are no university institutes. This leaves the universities with a too passive role.

Unfortunately, the new Union programmes for the support of academic research have been overloaded with criteria for the allocation of funds. Professional lobbyists will receive most of the funds and the universities will have a hard fight.

At the moment, mainly the US and Canada are participating in our exchange programmes. We should further extend them to other parts of the world, in particular to the countries of the CIS. Exchanges with these countries will probably be asymmetrical for the foreseeable future. Their value will be to help prevent a new division on our continent and to support the forces of modernisation.

This month, a conference will take place in Rotterdam on the topic of the "liberated university." The organisers would like the university to be freed from administrative and political constraints, most of all, and hopefully from some financial constraints. Universities should remain centres of study and education, in the forefront of intellectual endeavour. It is this type of university that we need most of all to educate future generations towards a united Europe. At the moment, we are still in the initial phase of this undertaking.

CONVERSION AS A TASK FOR THE ACTORS IN DEFENCE ENTERPRISES

MARTIN GRUNDMANN

This paper was presented to the working group "Mechanisms and Projects for the Restructuring of Industry in the Baltic Region and the Formation of New Markets."

I will concentrate on a single but important aspect of conversion: *the role of the actors of conversion within defence enterprises.*

The general conditions and the institutional framework for conversion are quite different when western and eastern countries in the Baltic Sea Region are being compared. Although these factors play an important role in the conversion process, they represent only one side of the problem. Despite the given conditions, conversion must be accomplished, together with the positive development and implementation of certain steps and strategies. I assign particular importance to the actors in defence enterprises, as it is their task to bring conversion forward in order to change the direction of business activities in their companies. Regarding the actors of conversion, specific similarities between Western and Eastern countries are evident. The conclusions drawn from specific German experiences will interest the Central and Eastern European countries as well as my presentation of two examples of conversion projects in trans-national co-operation.

1. Factors to Consider

Objective and subjective factors alike must be taken into account when discussing the sociological and psychological aspects of conversion. Actors occupying authoritative positions in their field of work relating to the conversion process (managers, works councillors, engineers), will behave in accordance with personal situational interpretation, having assessed civilian potential in defence technologies, facilities, products and properties, they will employ decision-making according to their valuations.[1]

For example, engineers often assign greater value to defence over civilian development, under the presumption that the technological level of defence goods is superior to that of civilian goods. These engineers are mainly interested in technological performance. Such an interpretation of civilian goods as below standard sets important restrictions on the freedom of civilian-oriented engineering creativity.[2]

K. Prunskienė and E. Altvater (eds.), Transformation, Co-operation, and Conversion, 95-101.

We are thus faced with the task of discovering the best possible practices or forms of assistance which would allow responsible actors to utilise their abilities in successfully dealing with conversion.

2. Traditional Views

Traditional business management, that is, the discovering of new products for new (civilian) markets and utilising their "core competences" and technologies in the design and sale of commercial products, has for a long time served as the centre of discussion in German defence companies - conversion has simply been regarded as a technical problem. It has recently become clear, however, that this technically-oriented and mechanistic way of thinking and dealing with the problem is no longer appropriate. Conversion does not solely encompass "switching over" from military to civil production, where highly qualified employees, technological standards, financial conditions and available markets are the only preconditions for success.

The traditional trade unionist way of thinking, that the transfer to civilian product manufacturing from defence goods may be accomplished by contracting the same employees to operate the same equipment at the same worksite under identical working conditions, is also more romantic than realistic. Many more changes must be implemented than has formerly been expected and anticipated.

3. Defence Company Culture

The rather "weak" factors within enterprises, such as organisational and communications structures and the patterns of employee behaviour (managers as well as engineers and workers), are those which play an important role in the conversion process and may hinder its initiation and forthcoming[3]:

- The conditions fostering arms production are rather gentle compared to commercial business. The arms industry market is protected, highly regulated and transparent, possessing few customers and competitors. Relations with customers are stabilised through long-term personal networks.
- The organisational structure of defence companies is highly differentiated, including many responsibilities, competences and, as a consequence of this, many formal executive jobs. This elaborate division of labour leads to a habitual way of thinking about departmental responsibilities that hinders problem-oriented work methods and innovative communication, which are necessary for conversion.

- Complete priority is given to the technical performance of products, defined by defence requirements. Thus the process of development is a very long one. The aspects of cost and time are, in principle, neglected.
- Opportunity for individual fulfilment on the job (and therefore job-satisfaction) is very high, especially in development departments.

The combination of these and other weak factors creates a syndrome which could be characterised by the term "defence company culture". And this defence company culture interferes with conversion efforts. It is necessary, therefore, to transform the defence company culture into a company culture which is appropriate for civilian or commercial markets.

FIGURE 1. *Defence Company Culture*

1. Compared to commercial business, the conditions of arms production are mild:
    - stabilised relations to the (only) customer, little competition
    - large periods of technical development
    - the lack of cost-consciousness is necessary (functional)
    - large scope of work
    - the technical play instinct may be cultivated
    - arms companies pay little attention to political and social development

2. Working in arms companies stimulates employee elitist convictions:
    - military technology means the best technological performance; this generates specific pride in producers.
    - the competitiveness of one's own products is extremely high; comparison with competing products is not realistic.
    - identification with the military customer is highly developed; relationships of personal trust have been established which substantiate successful work.

3. The organisational and communications structures in companies are dedicated to the processing of military orders:
    - regulations of secrecy hinder communication
    - formalistic routines restrict creativity
    - long-term specialisation on a single problem weakens flexibility
    - orientation toward pure technical problems hinders a realistic valuation of one's products

4. Arms production also influences civilian approaches:
    - the habitual way of thinking in departmental responsibilities and competences restricts problem-oriented work methods
    - anticipation of military orders restricts conversion activities
    - the presentation of commercial products is often unprofessional and not directed at (new) commercial customers

4. Required Organisational Changes

The mechanisms for changing this culture must take into account two aspects: organisational restructuring and employee retraining. Both aspects are interrelated and must be tackled mutually.

At least three areas in organisational structure require changes:

1. Civilian (or commercial) business needs organisational elements which the defence business lacks: controlling, marketing and a customer-oriented sales department. This is the most crucial point, since the establishment of these functions is a precondition for success in civilian markets. Most conversion projects fail because new civilian products do not meet customer and market demand. The products are often technically excellent and the enterprises are proud of their success in conversion - however, there are few buyers, as the differing consumer needs have not been systematically analysed in advance.
2. There are three reasons why commercial business in defence companies must be separated from the defence business and independently organised:
- commercial and defence business cultures are different - their separation provides for easier development of a commercial business culture;
- competition for company resources (employees, time, capacity) among commercial and defence orders should be avoided;
- the need for secrecy in defence orders restricts necessary technological innovation in the civilian field.
3. Formal hierarchies in enterprises should be levelled and "flat hierarchies" and team structures should be introduced; positions and competences must be redefined and the participation of the workforce in the conversion process intensified.

Development of substantial changes in the organisational structure of a company such as the above-described can only occur through a long-term process - but this process must be initiated as early as possible.

5. Reorientation of Employees

Orientation towards an organisational structure which has become highly adapted to arms production during decades also reflects normative power, and managers and employees have become accustomed to it. It is therefore necessary to take a closer look at those individuals, the actors in the companies, who are highly affected by organisational change.

Most defence companies employ a highly qualified workforce. They do however, face problems in adjusting to those civilian production requirements which companies consider subjective factors hindering conversion.

One of the key words is cost-orientation. When defence company employees develop civilian technology, they often retain habits and practices acquired in defence production. It follows that civilian products are developed according to the maxim "design to performance". But the rules of the game in civilian markets are different. Products must meet both consumer needs and optimal market prices. It is, therefore, necessary to change the maxim to "design to market" and "design to cost".

This change is regarded as a process requiring the "conversion of minds", the surrendering of defence-related habits and routines. It is not only necessary to make civilian use of defence-related technology but to utilise it in a way that meets civilian market conditions. Conversion is not only a technical and economic problem but also a social one.

A closer examination of this problem reveals that minds and attitudes are manifested in patterns of behaviour or, to use another term, in job roles that are adapted to defence production conditions. Therefore, it is not only necessary for management and employees to learn new functional skills such as marketing, but they also have to adjust defence-oriented behavioural patterns and job roles. Retraining is necessary for this task.

Retraining programmes often concentrate on new functional aspects of commercial production. But the normative aspects of work should also be taken into consideration, since job behaviour is closely linked to normative expectations.

New values must be defined to replace the old ones so that the fulfilment of new job roles becomes possible. Changes in behaviour, however, also require changes in the organisational structure of enterprises, because employees adapt to the work context through specific routines of job behaviour. Also, for this reason, a distinct division should be made between the defence and commercial businesses.

6. Western Assistance to Eastern Companies

Several examples serve as indications that defence companies in Central and Eastern Europe face similar problems, however, there is a lack of systematic research.[4] Moreover, compared to Western companies, Eastern companies are faced with an inferior economic environment. Nevertheless, they strive to adjust to market conditions and attract foreign investment. But foreign capital only finds its way when there is opportunity for profit. Western companies seeking Eastern partners expect market-oriented organisational structure and behaviour. It is therefore of priority for defence companies in Central and Eastern Europe to meet these expectations to some degree. Western assistance should concentrate more on this aspect of the conversion process in Central and Eastern Europe. Two examples of such initiatives follow.

1) The Swedish non-governmental organisation SWENESCO (Swedish Engineers, Economists and Scientists for Conversion) carries out conversion projects granted by the Swedish government, with companies in Central and Eastern Europe, for example in Tallinn or St. Petersburg. They establish co-operation networks between a Swedish and an Eastern university and Eastern defence companies and organise business contacts with Swedish companies. The project's main activities include analysis of problems in defence companies and the presentation of solutions in the areas of marketing, product pricing, organisation, corporate strategy, etc. Work is performed by Swedish students under the supervision and guidance of experienced scientists. The results are published as diploma works in which the valuation of the process by the Eastern company is also listed.[5] Intensive collaboration between all partners occurs throughout the entire process. Students from the Eastern partner university receive an opportunity to conduct their diploma work in Sweden. This initiative works on a very practical level and is not very expensive.
2) The German institute COGNOS International, in co-operation with the Polish Orgmasz Institute of Management and Organisation in Industry, has set up a retraining programme for higher management in the Polish defence industry. The project is granted by the German Ministry of Economics and is organised in both Poland and in Germany. In Germany, contacts between Polish firms and German companies with experience in conversion were made possible, and during a one-week seminar at the Management Academy in Bad Harzburg, an intensive retraining was carried out.

Both initiatives are positive examples of co-operation between East and West bringing conversion forward. They illustrate that conversion is not only a macro-economic problem but also a task for the actors in companies. From my point of view, co-operation between universities and companies is a very interesting means of assistance.

[1] Martin Grundmann, Margitta Matthies, Klaus Potthoff (1995) Rüstungskonversion: Erfolg durch Wandel der Unternehmenskultur (Arms Conversion: Success by Changing the Company Culture), Lit, Münster/Hamburg.

[2] Patricia L. MacCorquodale, et al (ed.) (1993) Engineers and Economic Conversion. From the Military to the Marketplace, Springer, New York.

[3] The largest German arms producer, Daimler-Benz Aerospace, also shared the same experiences, cf Dietmar Schrick, Praxiserfahrungen bei betrieblicher Konversion und Diversifikation (Experiences in the Practice of Conversion and Diversification), BWI-Wirtschaftsdienst 4, 21-26.

[4] Several contributions to the international conference Conversion in the Baltic Sea Region, organized at the University of Kiel in March 1995, dealt with this aspect, see: Martin Grundmann (ed.) (1995) Arms Conversion in the Baltic Sea Region and in Russia, Lit. Münster/Hamburg.

[5] Linköping University (1993) Provedenie konversii na zavode apparatury dal'nej svjazi, Linköping University (1994) Electron - Western Perspectives on a Russian Research Institute, Address: Swenesco, Linköping University, Department of Management and Economics, Dr. Rune Wigblad, S-58183 Linköping, Sweden.

CONVERSION POLICIES IN AN INTERNATIONAL PERSPECTIVE: GERMAN POLICIES IN THE BALTIC SEA REGION

ULRICH ALBRECHT

1. Introduction

The end of the East-West conflict caused a shift in the conversion issue from an academic and trade-unionist debate to actual affairs. Today conversion is a serious concern for local and regional policymakers, for headquarters of the arms industry and for their employees. By now the societies of Western and Eastern Europe have accumulated a certain amount of experience in arms industry conversion. Now, the learning stages and processes in the handling of the conversion issue, especially in the CIS and the former WTO member states, can be discerned.

There seems to be no consensus on the yardstick for measuring success in arms industry conversion. Trade unions refer to continued employment of the workforce. Managements stress successful diversification and company growth (or other managerial objectives). In more general terms, the performance of the once dominant defence sectors of former communist societies in the transition processes will be an issue of concern. From an even more general viewpoint, ecological and regional reconciliation aspects of various interests may be considered. It is thus necessary that the desired goals of the conversion process are made more clear, in order to define successful conversion and to review sectoral policies.

Despite the lack of commonly accepted standards, it may be said that the actual conversion experience is distinctly disappointing. The few examples of former defence combinates which operate successfully are far outnumbered by the large number of closures and bankruptcies. This general failure should spur debate on whether or not past theoretical efforts in conversion/transformation have been inherently inadequate and require improvement. Conversion is part and parcel in the all-embracing process of transition in Europe. The former WTO member states in Central and Eastern Europe are faced with a double task of converting their economies from semi-war economies to civilian ones (referring to the main contents of the production effort) and switching from command economies to market structures (this refers to the regulatory aspect of production). In Western Europe, ecological constraints and the employment problem enforce a similarly radical reorientation of economic strategies towards directed transition to a sustainable world.

K. Prunskienė and E. Altvater (eds.), Transformation, Co-operation, and Conversion, 103-117.

The conversion of the defence sector can politically be very suitable to provide the leading element in this general process of transition, since the public may possibly employ the argument that after disarmament some basic alternatives must be articulated and implemented. Such a course also entails potential for the further democratisation of the political process.

2. General Requirements for the Safeguarding of Transition

The radically changed security situation in Europe demands fresh initiatives to safeguard the great transition. The question is: which political framework will facilitate the process? Prime candidates for this requirement commonly exist in Western institutions, lead by NATO. These institutions have been moulded during the Cold War. The old NATO with its preoccupation for military preparedness faces unconventional challenges, and the question remains whether the dramatically changed situation deserves new institutions. There is a need to establish a politically secure environment, which could enshrine arms industry conversion. If, e.g., the Baltic Republics were part of the Western Alliance, the reorientation of the defence production capacities of the WTO and Soviet era would follow a distinct path of integration into Western European collaborative arms production schemes. In the absence of such a large political scheme, more moderate transitory steps could be sufficient. The present state of affairs, however, the NATO Co-operation Council and the Partnership for Peace Program, clearly represents surrogates. These are not part of the Treaty. The extensive literature on this topic is dominated by the question of widening the Alliance, through the inclusion of the Baltic Republics. Too little attention is devoted to the issue of whether the Treaty itself, after five decades, deserves reshaping. A duration of half a century is extraordinary, considering the appropriate lifetime of international institutions. After such a period, a ”bottom-up” review of the Alliance is appropriate and nothing discriminatory.

The new Europe must develop its own defence identity. The US could and should play a role in this, but not in the way NATO is constructed. The new threats to security are the failure of the East to master transition, with ensuing problems for stability, and the consequences, namely stagnation in economic and political development, large flows of refugees, continuing ecological threats. A new European security order must envisage these threats. The former WTO countries ought to be integrated into such concepts - and not associated into a Western European security pact.

This calls for a first German political task in the new era. The new unified Germany mainly looks in two directions - West and (now) East, whilst the northern and southern dimensions remain of lesser importance. The Eastern neighbours of the new Germany justly have high expectations in the contributions of this country to re-establish political structures, the recreation of the political Europe in the centre and the east of the continent.

3. The Issue

As recently as 1992, the "peace dividend" was given a prominent status (e.g. in UN-reports) and estimated at 1.2 trillion USD for the last decade of this century.[1] The desired linkage between disarmament and potential development spurred special hopes in the newly independent republics of rebuilding the economic foundations of society. In the few years which have elapsed since the great changes were introduced, the Russians are, above all, facing the disappointing reality that the intended conversion of the national military-industrial complex (VPK) made little of only insignificant contribution to the well-being of the population. The conversion literature about the societies, which contributed to the outrageous Soviet armament effort, e.g. the Baltic Republics, also largely mirrors these disappointments.

This state of affairs deserves considerable analysis of its causes. Military industries in the former Soviet Union have submitted a range of alternative product ideas for civilian markets. These include supersonic business aircraft, a number of utility and hobby aircraft in a sector which was formerly neglected in the USSR aerospace industry, heavy duty tracked equipment offered by former tankmakers and a spectrum of merchant ships designed by former naval yards, up to nuclear-propelled underwater tankers and freighters. Supporting industries wish to sell a number of durable consumer goods, from leisure electronics to basic household appliances such as refrigerators, dish washers and motorcycles. There is no reason to doubt the technical skills and abilities, especially of former defence firms, to develop alternative high-tech and medium-tech products for civilian markets.

Yet conversion is apparently understood in a much too narrow light, if it solely focuses on alternative products. These new products offered to civilian consumers deserve detailed evaluation, and in a number of instances they appear wrongly perceived, badly designed, with poor customer appeal. But the fact that they largely failed to attract customers on the domestic market of abroad is apparently to be explained by non-technical factors, namely existing structures of manufacturing and distribution.

The assertion that producers of military equipment generally lack sufficient potential in the arts of marketing, after-sales servicing and support or helping potential customers in organising financial credits, will not be appropriate, but it is also not the root cause for the failure of management in the former VPK to organise a smooth transition into commercial life. The actual causes appear as more comprehensive in nature, encompassing more tactics and the design of attractive products.

This hypothesis is substantiated if one looks at the parallel situation in China. Circumstances are comparable; before the conversion of arms-making capacities in 1985-87, China also had an over-extended defence industry, accompanied by huge excess demand for durable consumer goods such as bicycles, washing machines, refrigerators and leisure electronics. Large-scale imports of such items were not possible. The conversion of part of the defence industry to the manufacturing of such goods apparently proved to be a major success in China.

The difference in the situation of the former Soviet states is not apparent in the realm of technology. The main difference from the Chinese situation is that ex-Soviet societies are involved in an all-embracing transformation effort, reshaping one-party political systems into democracies, and converting planned economic systems into market economies. Thus the causes of the lack of success in arms industrial conversion ought to be sought in the sphere of industrial organisation and its restructuring. Political leaders in government and industry apparently had greatly underrated the structural economic devastation inflicted by the arms race on industrial structures and the special requirements to develop alternatives.

The combination of an obvious over-militarisation of the economy and the systemic characteristics of a "permanent Kriegswirtschaft (war economy)"[2] as the point of departure made successful conversion a concomitant (pre)condition of transformation of society as a whole. Under these circumstances, it is not surprising that conversion was at the top of most agendas dealing with the stabilisation and transformation of ex-Soviet countries. They have been executed with little success. Resources continue to be wasted on maintaining industrial structures that are bound to close down in a market economy, while necessary investments for modernisation cannot be brought to fruition. Many crash programmes were suggested, few were launched, hardly any were turned into success. However, the governments have so far managed to avoid a total collapse of the economies, in spite of gloomy predictions by a number of academic economists.

Actual conversion problems which are typical for the situation occur in such environments. If a military plant or facility aims at producing civilian commodities, it remains restricted to draw upon in-house upstream supplies. A resulting product is likely to reflect technical features defined by the old-fashioned military conditioned upstream production capacities in contrast to production styles created by market-oriented production. The outcome of the still dominant military-industrial setting is presently typical for the so-called civilian conversion products which prove to have meagre consumer appeal. Examples are provided by the exhibits at present Russian "Conversion fairs," from "Conversiya '90" in Munich, Germany, in April, 1990, to the so-called BDNX in Moscow in Spring, 1995. There one finds the vacuum cleaner an average person is not capable of lifting, or the multi-purpose agricultural tractor which does not meet basic ergonomic criteria and which is simply not needed in a generally over-saturated market.

Counter-strategies should be two-fold: to advocate the responsible figures to further dismember the huge military production conglomerates into a myriad of smaller, market-oriented enterprises, in order to break up former military-induced production styles, which are particularly alien to the small-scale structures in the

Baltic Republics. Secondly, the kind of product offered to customers needs radical changes, if conversion is to be a success. The product should reflect the demands of the market rather than the past of the maker.

For the time being, it seems, a majority of military factories in the former USSR has adopted a "defensive" position where available funds[3] are used to maintain the workforce at minimum wage levels[4] and often reduced working hours. Sometimes military production is continued in Russia even in the absence of government orders. Presently, a preferred scheme to find a rapid way out of the morass is to separate high-tech sections from main enterprises and to offer these, looking for Western partners, in order to create new privatised companies.[5]

Additionally, well-intentioned Western observers tend to overlook the fact that any transition strategy requires a realistic assessment of the status quo ante because even the most radical industrial policy will be determined by the parameters of historical continuity. Seasoned by the old system, the industrial leadership of the military-industrial complex managed to exploit existing uncertainties.[6] The old networks formed a cohesive web of interest groups which learned to adapt and to survive in the new setting of postulated transition as well as to improve their personal situation (wage differentiation within enterprises has been considerably increased, favouring the top layers of management in particular). The Russian Government felt forced to concede such developments and to legitimise ex-post substantial amounts of finances created "subversely" through reciprocative inter-company credits. However, not all enterprises and research establishments closely associated with the VPK are equally well situated and connected in order to muddle through and withstand the pressures for rapid restructuring and conversion.

In Western countries, the wake of reduced arms procurement due to international disarmament accords is also not usually characterised by conversion. Companies often respond by sizing-down their branches by means of acquisition. Thus managements in former Soviet countries, which look to Western patterns in order to convert to market economies, fail to get positive demonstrations in exactly the field where they need it most - conversion of arms plants on a large scale.

The optimism of the early years of perestroika has faded, but it remains doubtful as to whether an unrelenting search for the causes of the failure of conversion efforts in ex-Soviet countries has begun. "Conversion" no longer figures prominently in political debate. Among the prevailing reasons contributing to failure and even resistance to conversion measures, the inherited industrial organisation contributes substantially. Institutionally, production sites and R&D facilities, for instance (with their innovative potential), were kept separate under the old system. The high-tech elements of the former Soviet armaments sector, as they are represented by R&D establishments, especially appear threatened by failure in contrast to enterprises oriented toward production in fields of medium technologies, which find customers in domestic markets. The present orientation of R&D establishments toward self-financing civilianisation appears additionally hampered by the distance to production. A recent survey of high-tech industries in

St. Petersburg (cf. section 5.1) reveals profound differences between the Russian approach to organise R&D and the corresponding Western structures. In the West, R&D facilities are commonly incorporated into the overall structures of companies, also in order to exploit their innovative potential as much as possible.

It remains difficult to imagine that the large ex-Soviet R&D establishments can successfully convert to commercially viable units on their own and without state subsidies. The determination of management and staff to survive are impressive but will not be enough to keep them alive as separate large entities. Western experience highlights the general difficulties of organisations operating within the military realm to supply technological spin-off to the civilian economy.[7] In general, the "military culture," at least when it dominated to the extent it did in the former USSR, is not conducive to the productive flows of knowledge to the civilian sector.

If nothing more happens, especially the large R&D facilities in the countries of the former USSR face elimination, or at least dequalification to non-high-tech work. This would result in tremendous loss in qualified human capital for the societies involved. Help from outside, granted specifically to address pivotal points in the conversion effort, could act as a surrogate to national means of re-orientation, which do not exist.

4. Present Programmes

Several international programmes in support of conversion and transformation of the CIS economies are being conducted locally, also in the St. Petersburg region. Most often business seminars (with simultaneous translation) are being offered, in which specifics of the defence industry are more or less treated in detail.

A second kind of conversion support measure is provided by a number of Western countries, which offer traineeships for CIS nationals in the industry and service sectors. These commonly require extensive and costly language training, which can be considered an additional asset. It was found that especially in the CIS defence sectors, a great number of the more qualified employees command English to some extend (as was convincingly shown, since they understood articles in Western technical journals and publications).

Similar examples are the offers by the Danish Government for business classes, including a period of practical training in Denmark. There appears to be an especially large number of programmes in Germany to support CIS transformation efforts, more recently attracting managers from the defence sector in greater numbers. The Government of Baden-Wurttemberg sponsors a business training centre offering a variety of courses, some of which culminate in a three-month practical training in Germany. Other Lander (regional states) in Germany have developed comparable programmes, mostly supporting one respective training centre on their territory, inviting CIS nationals with information material translated into their native languages. The Deutsche Management Akademie Niedersachsen, the parallel in Lower Saxony, charges 50% of the costs of these courses to CIS participants.

A number of training centres set up to qualify business managers of countries with formerly state-planned economies, have integrated conversion into their educational programmes in market economics. One example is provided by the Bildungszentrum am Mueggelsee ISVG mbH in Berlin, Germany. Since its inauguration three years ago, this centre has trained more that 500 participants from CIS countries. In 1992, the centre started courses in conversion for top managers of the arms industry. Demand for these courses proved to be so strong, that a number of parallel seminars were established.

The head of the centre stresses that his students from the CIS countries are more interested in practical experiences in the transition which has progressed in the former German Democratic Republic, and that they disregard theoretical qualification. The objective of his centre is not only to train managers, but also to revitalise, answering the demands of his students, economic ties between Germany and Russia.

Some of the more expensive programmes carried out outside the CIS contribute to the further formation of persons whose qualifications are eagerly sought by foreign companies, which aim to enter the Russian market and which can afford high salaries. The patterns of recruitment for such future industrial leaders are presently not very obvious. Even though candidates for such courses were selected or mandated by large Russian enterprises, they are not likely to return to their former firms for a long time.

Along with training programmes for managers, a number of more comprehensive schemes for conversion emerged, combining the research with the training element and aiming at international collaboration. The following contains a brief survey of four international conversion training and research projects. The selection does not claim any sort of representation. The criterion for taking these classes was significance for Western collaboration, especially from Germany. These modest schemes, after the failure of large-scale programmes, may offer the potential needed to reset the gargantuous structures inherited from the past into lively and sustainable smaller units.

The UN University intends to set up a research and training facility for conversion along these lines in St. Petersburg, to be administered in the beginning by WIDER/Helsinki, which should also serve the transformation interests in the Baltic Republics.[8]

4.1. HCA PROJECT TO ACCOMPANY THE TRANSITION OF THE ZTS MARTIN TANK PLANT, SLOVAKIA

Mary Kaldor, co-chairperson of the Helsinki Citizens' Assembly (HCA) with a long-standing record in conversion in the UK, pioneered a scheme by which a group of Western conversion analysts monitors the efforts of a major tank factory in the industrially less developed new state of Slovakia. The involvement of the

HCA in such a specific project requires explanation. The newly democratic Czech Republic inherited, after the "Velvet Revolution," inter alia the large ZTS Martin plant, a major WTO resource for producing tanks with some 50,000 employees, in an otherwise touristic environment in Eastern Slovakia. The politicians in charge after the revolution, especially President Vaclav Havel and Foreign Minister Jiri Dienstbier, stated repeatedly after taking office that democratic Czechoslovakia would not export weapons any longer, and that the industry should convert. After learning about the huge tank plant in economically underdeveloped Slovakia, President Havel turned to the EC and received a generous half-year grant due to the personal involvement of Mr. Jacques Delors. When the EC eurocrats sent out from Brussels ran into difficulties to come to grips with the Martin situation, disgusted managers of the plant called on the HCA for help.

The HCA assigned one of the conveners, i.e. the present author, with the responsibility for "Conversion and Ecology," to come in with a helping hand. A number of social scientists, headed by Mary Kaldor, agreed to pay regular visits to ZTS Martin and to discuss current problems in a seminar-style arrangement. The (US-based) Institute for East-West Studies (formerly the Institute for East-West Security Studies) provided financial support, via its Prague-based outpost, and participated in the evaluation of conversion efforts.

The conversion concept of the ZTS Martin management was soon found to be very conventional. Like other former tankmakers, ZTS opted for agricultural tractors, both of the wheeled and tracked varieties, hoping for exports to Western (hard currency) countries. A cross-check with leading Western tractor firms rapidly confirmed that there are no special export opportunities for Eastern European competitors, especially because of their weakness in post-sale support to customers. ZTS also bided successfully because of low wage costs, in answer to Western tenders, such as metal work for casings for the London underground transport system.

HCA activities included a delegation of economic development experts to Slovakia in September, 1992, a small delegation from Slovakia to the UK in December, 1992, and another visit of experts in March, 1993. One staff member of HCA Prague is going to move to Martin and take office in the ZTS headquarters. A "Foundation for East-West Development Initiatives" (FEWDI) was established at Martin in September, 1993, and some progress has been made towards initial funding to set up an office for FEWDI. In 1995, it was decided to place an employee of a leading US consulting firm at Martin on behalf of the HCA.

Presently, the effort of the HCA accompanying group is geared to educate local ZTS management in the proper drafting of applications for grants (instead of turning to commercial banks, the management, for instance, called upon the government to facilitate credits in the manner customary before the revolution) and to familiarise them with more advanced approaches to the conversion issue. Besides the chairperson and the convenor, one staff member (out of a total five on the Prague HCA staff) is assigned to support conversion activities on a permanent basis. A general advantage of the HCA network is its ability to recruit fresh Eastern European talent.

4.2. BICC

Very closely related to the UNU project proposal is the foundation by the German regional government of North-Rhine Westphalia, Germany, of the "Bonn International Centre for Conversion" (BICC).[9] The purpose is "to promote the process by which people, skills, technology, equipment and facilities in defence are shifted to alternative economic applications. The Centre will focus on actions that make the process of conversion more effective, while its services are intended to minimise hardships resulting from disarmament." This new centre began operating in spring, 1994. The regional government hopes "for participation of other organisations."[10] The Bonn Centre is also likely to submit, in collaboration with the UNCTAD Programme on Transnational Corporations, a proposal for "technical assistance for projects in Russia and other countries in transition."

The founding document of the BICC defines its task in more detail: "to service, on the basis of practical experience and scientific and empirical knowledge, the management of disarmament and to assist in practical conversion and demobilisation projects." This new centre will focus on six areas:

- Reallocation of financial resources used for military purposes to foster disarmament and conversion;
- Reorientation of military R&D programmes towards major global challenges, underdevelopment and the degradation of the environment;
- Industrial conversion;
- Reintegration of military personnel and civilian personnel employed by the armed forces into non-military jobs;
- Reallocation of military facilities and installations;
- Alternative use or scrapping of surplus weapons.

Recently BICC has also turned to the conversion problem in Lithuania.[11]

4.3. THE PROGRAMME BERLIN, ST. PETERSBURG

Conversion troubles in arms plants in St. Petersburg induced a scheme of co-operation among these firms, Russian social scientists and Berlin universities. Because the Russian firms were unable to afford the high fees of Western economic consultancies, they approached the author with the request for respective support within technical departments of Western universities. The transition of a shop manufacturing armoured personnel carriers toward a facility producing vacuum cleaners was supported by a German MA candidate in economics of manufacturing. The private German Berghof-Foundation facilitated a two-week on-the-job training opportunity for this individual in the Bosch company and financed the travel cost to Russia. In St. Petersburg the German MA candidate conducted a four-week study on the possible improvements in the production of vacuum cleaners. His hosts paid the local costs for the stay. Russian social scientists came in, originally to help in translating, later to accompany and support the Russian-German project. The project report with findings of the young German expert were submitted both in Russian and German. This scheme was developed into a model; counterparts in St. Petersburg submit a specific request for expertise in a conversion programme, mostly durable consumer goods, which is disseminated in academia in Berlin. A suitable candidate is nominated, sometimes opting to work for a thesis (as is also the case in the SWENESCO scheme, see 4.4.). The individual is supported by an academic supervisor in professional matters, and by the Berghof-Foundation with costs incurred by the project. Local costs are the responsibility of the hosting conversion partner. The findings are made available to all parties involved, including the local government. Experiences of participants are disseminated by a circular letter (presently five issues), which is available on request.

The collaborating Russian and German social scientists have formed a formal unit in the meantime, the "European Association for Research into Transformation," also in order to facilitate communication and to save fees and other dues in the new republics.[12]

4.4. THE SWENESCO PROGRAMME

The basic insight that successful conversion in formerly state-socialist countries requires adequate academic qualification of cadres, has spurred the creation of a number of ad-hoc schemes between Ex-WTO military plants and Western academic institutions. Another reason for this resort to academia is the mere fact that enterprises in the former Soviet Union simply cannot afford the fees for Western commercial consulting firms, and sought (and found in a number of cases) direct contact to academic institutions in other countries besides Germany.

One of the first activities became known under the name of SWENESCO (Swedish Network of Engineers for Conversion). This Swedish subsidiary to the International Network of Engineers and Scientists (INES), according to the chairperson, has the objective "to carry out a few pilot projects on the conversion from military to civilian production in some areas of the former Soviet Union." The budget for 1993 comprised about SKR 0.4 million, including money from Swedish sources and the MacArthur Foundation. As a first specific project, SWENESCO sponsored the academic support of conversion of a major shipbuilding enterprise, Shipyard No. 7 in Tallinn, Estonia. Four MA candidates for an engineering degree were sent to Tallinn to help the shipyard and supported by the Lund Institute of Technology, in collaboration with a Polytechnic in Tallinn.[13]The basic feature of the so-called SWENESCO scheme is worth recollection: a converting former arms plant in one of the ex-WTO countries collaborates with two academic centres, one neighbouring the site, giving general support, and a Western institution which fills in specific expertise or fresh talent. An international conference is convened, recruiting from the INES network, to assess the project. Among the participants are also representatives of regional professional associations (in the case of Shipyard No. 7, the new Estonian Technical Societies, ETS).

The next SWENESCO initiative is selected to be a Russian plant, previously specialising in equipment for the former Soviet submarine fleet, the electronic factory PERLC in Pskov (a town south-west of St. Petersburg, Russia).

5. Evaluation of Existing Institutions and Programmes

The findings of this study suggest a number of preliminary conclusions:

1. There is an array of research and training efforts aimed to support conversion in former centrally-planned economies, differing in scope, kind of activity, institutional set-up and programme. Cross-information about individual activities is sparse within this community. In addition, new institutions with more comprehensive programmes such as the German BICC are emerging.
2. The prime demand from former Soviet-dominated countries is the acquisition of practical experience. Conversion research is definitely considered to be of secondary priority, as well as theoretical qualification. None of the shorter seminar-type programmes received a particularly positive rating in interviews conducted with participants. The head of the conversion department in the St. Petersburg City Council pointed out that the training courses themselves were generally of little practical value, while practical training abroad would lead to regular business contacts with a potential for expansion.
3. Industrial conversion dominates the various programmes. The reorientation of R&D capacity ranks second in these activities. Other conversion areas, such as the alternative use of existing stocks, the reintegration of military personnel into the civilian economy, the scrapping of surplus weapons, or the reallocation of military facilities, are apparently not considered to provide long-term tasks.
4. Emphasis on existing conversion programmes related to ex-Soviet countries rests with the areas of business economics and business administration. Appropriate conversion products or production technology are considered to be areas of prudence where no or limited outside support is needed.
5. Whilst the majority of existing activities appear to be mere training programmes, projects directly related to ongoing conversion efforts (HCA/WIDER, SWENESCO, the Berlin activities) appear to contain the greatest potential for actually steering the reshaping of former centrally-planned economies and avoiding bankruptcies of armaments plants.
6. These more adequate activities regularly involve academic support, provided both by domestic and foreign institutions, which co-operate on a bilateral basis in the conversion activity. Most of the consulting work is carried out by

graduate students who write their thesis (SWENESCO, Berlin project) under proper supervision. Commercial consulting firms play a limited role, compared with Western conversion activities.

7. A feature of a number of active programmes is their international composition, including foreigners leading the counterpart personnel (HCA/WIDER, SWENESCO, South Korean project). Such mixtures apparently broaden the body of experience available for the conversion effort and are seen as a contribution to widen the base for future economic ties.
8. Apparently no programme has so far reached the upper echelons of the large military enterprises. Top management, representing the real assets of the old system, does not perceive itself as potential pupils or recipients of advice. In conjunction with a systematic overrating of the market value of their assets, they fail to develop a realistic approach to international co-operation and the required steps of restructuring in order to enter new high-tech markets.

6. Preliminary Findings

Disarmament and the drawdown of military activities require, for social and economic reasons, intensive and qualified management over extended periods of time. Practical and training assistance is required to ease the adjustment process. The most pressing requirements lie in the "soft" fields needed for the transition to a market economy: business administration techniques such as quality control, pricing and cost calculations and marketing skills. Disseminating experiences gained at one local centre is also considered important. The SWENESCO programme has the following overarching goal: "The gained experience will be documented and disseminated in order to stimulate and influence the 'market process.'"

Under the prevailing circumstances of support for transformation in the former Soviet Union, conversion measures should concentrate on creating facilities to train management in alternative structures of production and services which would wean qualified personnel from the military- industrial complex or, positively, allow these enterprises, under insightful guidance, to shed part of their workforce. The adequate preparation of large numbers of persons, with different levels of qualification, to work in market-oriented private enterprises, would make an important contribution to the transformation of the economy and help to convert the VPK.

6.1. THE STRATEGIC CONTEXT

Any concept of converting the colossal ex-Soviet VPK presupposes a vision of the future defence industrial sector in those countries, in order to recommend rational decisions in which capacities should be maintained and modernised and in which capacities are either converted towards civilian activities or closed down. It presently remains difficult to estimate the future defence posture of the new states and the implications for industrial policy from the outside. A feature of future defence industrial activities in these countries will be openness towards Western activities and collaboration with Western companies. Thus basic parameters of reorientation of Western defence industries will have an impact on their new counterparts.

The emerging industrial policy paradigm of future defence bases takes into account that advanced technology for military purposes has to rely on technology and innovation created within non-military global industrial networks to an increasing extent. As a consequence, Western defence industries are under major pressure to redefine their basic design philosophies and to restructure in order to be competent to continuously absorb innovation as it becomes available in world-wide civilian markets. Modulation and dual-use are the catch words of this evolution. The beginnings of these strategies, which are closely interwoven, rest in the concept to overcome export controls by explicitly going for dual-use characteristics of components of advanced weapon systems, particularly in Western Europe. Another motive to adopt this concept may be seen in the desire of the military lobby and the large R&D labs to gain legitimisation for future government funds for their support, in a period of disarmament and tight public budgets. Actually, dual-use refers to the incorporation of civilian innovation into complex weapon systems at increasing scales.

A debate about the real meaning of modulation and dual-use technologies is going on, but it remains important to consider the key defence industrial paradigm when advocating decisions on the future of the armaments industrial sector in the former Soviet Union. Accomplishments of the past may not be a warranted indicator to gear the sector towards the challenges of the 21st century. The actual challenge is apparently to mould the high-tech armaments sector of the new states into the globally evolving industrial networks, i.e. to open it to the world market. This indeed suggests strong international ties, something unknown to the VPK of the past.

Structural changes within the world market for high-tech items at large are to be selected to orient strategies of conversion and transformation, especially in the defence sector. Only an efficient organisation towards such ends is likely to produce the politically desired results. A policy which simply would shrink the VPK down to the proportions needed for domestic defence purposes, plus a modest export potential, would most certainly end up in the petrifaction of existing structures, with no or, more likely, counterproductive results for the overall transformation effort. In addition, even the biggest military producer, the United States, has begun to accept the reality of global economic competition also in the defence production sector, once the arcanum of self-sufficiency.

Hence, the new independent republics will only be able to maintain their defence industrial base as a technology asset if - and only if - this sector is placed in the context of an industrial structure, which must move rapidly towards full integration into the emerging global industrial networks. The high level of technological competence, which has been a match for the world leader in military technology, the US, can be preserved only through energetic integration into international markets.

The years since the great change have clearly demonstrated that the managements in charge of the large enterprises within the VPK behave primarily as political gamblers and only in second place as business managers. Interviews indicate an inclination of the more impressive managers to shed large parts of the workforce in order to concentrate on "lean production" largely for export, as soon as the legal situation will permit "management buy-out" privatisation.[14]The role of the old enterprises inherited from the centralisation of the past must be challenged from the outside.[15]If this evaluation is correct, supporting conversion in the former Soviet Union requires support for tendencies to de-centralise by advocating the break-down of conglomerates into smaller organisational units.

For higher management, it apparently remains a prime task to make them understand that enterprises with thousands of employees, a high degree of vertical integration, and separate R&D facilities, are not well prepared to enter conversion programmes. The famed "market orientation" could be implemented better through a radical deviation from the industrial structures of the past.

References

[1] United Nations, Human Development Report 1992, New York/Oxford (Oxford University Press) 1992, p. 86.

[2] Jacques Sapir, Les fluctuations economiques en URSS 1941-1985, Paris (Editions de l'Ecole des Hautes Etudes en Sciences Sociales) 1989.

[3] Among the sources of these funds, the deviation of investment allocations provided by the government and the income from sales of real estate as well as sales of stocks of semi-manufacturers and raw materials figure prominently.

[4] Such minimum levels often become supplemented by payments in kind earned through barter transactions, or a company organising wholesale purchases of consumer articles on behalf of the employees.

[5] In the absence of clear legal rules, such moves require the collusion of, among others, the state bureaucracy, thus opening all gateways for corruption. This provides for another operational link with a UN institution, the World Bank, and its programme to combat corruption, via a special institution, Transparency International (in which the present author is in charge of corruption in the arms business).

[6] In the old system of planned economies, withdrawing additional government credit above the limits set by the plan was punished rigorously with an imposed interest rate of 30%. The inertia of the bureaucracy which stuck to the old rules provided the VPK with a source of highly subsidized credits as the monthly rate of inflation approached this percentage (hint by Dr. Ksenia Gonchar, Moscow).

[7] US Congress, Office of Technology Assessment, Building Future Security, Washington, DC (GPO), June, 1992. The European experience is discussed in: Ulrich Albrecht, Die Nutzung und der Nutzen militärischer FuE-Ergebnisse für zivile Anwendungen (The Use of Military R&D Results for Civilian Application), Cologne, Germany (TÜV Rheinland: Technology Transfer Series, vol. 14) 1989.

[8] The Feasibility Study for this UNU RTC is available from the present author.

[9] See the "Conceptual Paper," prepared by (my long-term team colleague) Dr. Herbert Wulf for the Ministry of Science and Research, Foundation of the international Centre for Conversion, Bonn, Dusseldorf, Hamburg, June, 1993.

[10] Ibid., p. 2.

[11] Gediminas Rainys, "Economic Reform and Defence Industry Conversion in Lithuania," Bonn International Centre for Conversion, Report 3, Conversion of the Defence Industry in Russia and Eastern Europe, Bonn, 1995 (obtainable from BICC, An der Elisabethkirche 25, D-53113 Bonn).

[12] Contact: Dr. Peter Lock/eART, Auf der Koppel 40, D-22399 Hamburg, Tel./Fax 49-40-602 7972, e-mail:eartpl@oln.apc.org

[13] The MA thesis of the first two candidates is obtainable from the Lund Institute of Technology, Department of industrial Engineering (Magnus Larsson, Hans Nilsson, Conversion of Shipyard No. 7 in Tallinn, Estonia - not only a military/civilian conversion, Lund, 1992).

[14] Interviews conducted by Dr. Ksenia Gonchar and Dr. Yevgeny Kuznetsov in the so-called Northwestern Region (near St. Petersburg). Credits and thanks to these two Russian colleagues for this section.

[15] The famed Kirovsky Zavod (in Czarist times the even more famous Putilov works, a main scene of the October Revolution) in St. Petersburg - 50,000 plus employees, forms an outstanding example. In order to save viable parts of this giant enterprise, the firm has been subdivided into dozens of smaller entities, which still represent formidable firms.

TRANSFORMATION AND THE WORLD MARKET. WESTERN INTEGRATION AND EASTERN TRANSITION

ELMAR ALTVATER

1 Introduction

It is a trivial statement that the transition to a market economy, a pluralistic society and political democracy is taking place in the environment of the world system. The consequences of this statement, however, are not at all trivial. For it means that Central and Eastern European economies (CEECs) have to adjust their economies to the functional prerequisites of the world market (dominated by the highly industrialised OECD world) and within a short period of time have to fit into an institutional frame-work which has developed over years, if not decades. Therefore, it makes no sense to regard transformation processes as purely national ones, taking place within the territorial limits of either an old (e.g. Poland, Hungary, Rumania etc.) or a newly formed nation state (e.g. the Baltic States, the Czech Republic or Slovakia). The nation states and the political classes ruling them can foster the processes of transition, but they have to be aware of the "conditionality" exerted implicitly by world market structures and explicitly by world market institutions such as the World Bank, the International Monetary Fund or the institutions of the European Union.

This is the reason why transition processes have to be understood as taking place in an "interdependent world". Therefore they also require measures of adjustment to the new global situation after 1989 in the Western and even Southern world. These adjustments are not as extreme as the transformations in the East, but they have an impact on the economic, social and political development which has only in recent times become obvious after the first euphoric years. The transition, therefore, is much more than a pure transition from one state of affairs (central planning economy) to another state of affairs (free market economy): it is a transformation, i.e. the emergence of new social forms, a process of articulation of social forms possibly resulting in new hybrid-forms of social regulation.

K. Prunskienė and E. Altvater (eds.), Transformation, Co-operation, and Conversion, 119-127.

2 The Framework of the World Market

Therefore, in order to understand transition and transformation processes in the East, it is useful to briefly discuss the main features of the contemporary world market. With regard to Eastern Europe, two characteristics are of utmost importance. The first concerns the volatility of financial markets and the competition of currencies. Capital markets and their development are decisive for the economic trajectory in different parts of the world since their deregulation and liberalisation. The formation of interest rates and exchange rates takes place on global capital and currency markets; even powerful central banks cannot effectively control interest and exchange rate fluctuations. They have to respond to global monetary constraints with a monetary policy of internal and external stabilisation. Strengthening the own currency has the highest priority in all countries, not only in transition economies. Only by creating attractive conditions for capital investment will private capital flow into Central and Eastern European States. Three main factors influence the attractiveness of an economy: firstly, a low currency and country risk compared to other currencies and countries, secondly, positive interest differentials, and thirdly, low unit labour costs, i.e. positive effects on either wage, working hours, labour intensity or/and productivity as determining factors of profitability of capital.

The first two conditions are purely monetary ones. The third, however, emanates from social relations (industrial relations). It is also a consequence of material and institutional factors influencing structural or "systemic competitiveness" of a given economy. Moreover, the three factors influence each other. In the case of high systemic competitiveness, the currency risk is low so that the interest differentials with regard to "strong currencies" might also be low. This constellation is a benign prerequisite for investment which, carefully undertaken, again improves systemic competitiveness. In the opposite case, a vicious circle could also be triggered: low systemic competitiveness with regard to competitors on the world market results in higher currency risks which require high interest differentials and, therefore, hamper real investment into productive capital, create unemployment with again negative effects on industrial relations and competitiveness.

Therefore, economic policy has to follow two combined and related paths. It must stabilise the currency and equally improve the material and social conditions of competitiveness. It has been shown that this double task is easier fulfilled under liberal conditions than under conditions of centralised decision making (Transition. The Newsletter about Reforming Economies, Transition Economics Division, The World Bank, Vol. 6, 11/12, p. 4-6). But it should also be clear that liberalisation cannot successfully be implemented when released. Since stable and co-operative social relations are of utmost importance for systemic competitiveness, liberalisation policies have to take into account social interests of different interest groups and social strata. After a "first wave" of shock-therapeutical proposals, political institutions as well as scientific circles are well aware that the transition to a free market economy, i.e. the integration into the world market, must be socially moderated.

The second feature of the contemporary world market is the formation of regional integration areas. After the "first wave" of regional block formation in the 60s and the "lost decade" of the 80s, characterised by the debt crisis and its social and political repercussions almost throughout the so-called Third World, a "second wave" of regional integration areas - from NAFTA in North America and Mercosur in the southern cone of Latin America to AFTA in South East Asia, APEC in the Pacific Rim and SADEC in Southern Africa - has been triggered at the beginning of the 90s, in particular by the undoubted success of Western European integration after the Single European Act (1986), the Maastricht treaty (1991) and the transition to a Common Market (1993) with the perspective of forming a Currency Union before the end of the century. Obviously, the world market with its free trade discourse following the rules of the recently established WTO is "too big" for single nation states, whereas the traditional, even the large and powerful nation states are "too small" to perform well and unproblematically in the global competition. This is the main reason why intermediate organisations, regional integration areas (RIAs) emerge. It could happen that RIAs become areas of influence of the leading "triad" powers: the Western Hemisphere organised by the "Enterprise of the Americas Initiative" of 1990 and held together by the promises of forming a Free Trade Area from Tierra del Fuego to Alaska which is dominated by the USA, the Pacific rim hegemonized by Japan and possibly China in the future; the European Union becomes not only the centre of gravity but of economic and political hegemony with regard to the Mediterranean region and CEE.

What do these tendencies mean for economies in transition? Without any doubt, they also have to follow the trends of regionalisation of the world market, either by forming new regional integration areas or by joining already existing integration schemes. In the case of the CEECs, many of them have meanwhile applied for membership in the European Union (Poland, Czech Republic, Hungary, Slovakia, Slovenia), although they formed the Central European Free Trade Agreement (CEFTA) in 1994. The so-called Visegrad group mandates that all trade barriers on industrial products should be removed by the year 2001. But since trade with Western European and other OECD countries is much more important than trade among CEECs, the option of becoming a member of the EU is quite attractive. Other countries beyond the Visegrad group (e.g. the Baltic States) try to find some mode of association with the EU, and others again try to establish schemes of regional co-operation of their own (Central Asian States). These tendencies make clear that free trade is nothing more than a myth and that politically regulated trade regimes do matter. Therefore, rather weak states cannot avoid seeking a place in a trading block.

For CEECs the European Union is of utmost importance, because large amounts of foreign trade of CEECs is carried out with EU-countries. Most CEECs have high current account deficits and some even have a negative trade balance, as the following table shows. The largest part of the deficits stems from trade with Western Europe:

TABLE 1. Current account balance of CEECs

Current acct. balance	1993	1994	1995	1996	1997
CEECs and FSU	-5,1	-5,3	-3,6	-6,6	-12,5
6 CEECs*	-8,5	-4,4	-6,2	-7,4	-8,6

Poland, Hungary, Czech Republic, Slovakia, Romania, Bulgaria 1995-97 estimates. Source: OECD

The table contains data covering the period from 1993 to 1997. The projections show that until the end of the decade, the current account deficit is structural. In the cases of the Visegrad-CEECs, the current account deficit is a consequence of a negative trade balance, which in the case of the FSU, however, is rather favourable. The negative trade balance is due to protection of sensitive industries of the EU from emerging competitors in Eastern Europe, such as in steel markets, in agriculture, chemicals, textiles and machinery. But this is not the only reason for current account deficits. The more important fact is the lagging competitiveness in many industries where "world class production" is required. The standards for surviving on highly competitive markets are set in the leading economies of Europe, North America and East Asia. The EU is attempting to strengthen its competitive position with regard to the "triad economies"; the strategies of achieving this goal are comprehensively laid down in the White Book of 1993. However, by gaining competence in competition with the highly developed economies of East Asia and North America, competitive advantages with regard to Eastern and Central European Countries also increase. The consequence of this deliberation is rather disillusioning. Countries in transition will only be competitive in traditional industries where EU markets are highly protective, or they still have to define those "niches" where they might successfully compete with industries from the highly developed countries. These niches are possible, since modern technologies and globalisation of economic processes in general have lead to the strategic opportunities of splitting up the surplus value chain. Parts of this chain may be localised in transition economies, where conditions for optimising strategies of TNCs are good. The problem, however, is that the links of the chain are fully integrated into TNC-strategies, and that regional forward and backward "linkages" with other industries in a given area are rather poor. Therefore, the likelihood of positive economic and social development effects in a given region or country is rather small. This is a lesson which has been taught in a large number of cases in developing countries, which have founded their development strategy on the import of foreign capital.

Anyway, it needs some time of trial and error in order to find an adequate niche on the world market. Yet this time does not exist, because most transition economies are externally highly indebted so that they have to create immediate currency surpluses in order to transfer debt service abroad. Here again (as a couple of years earlier in the case of highly indebted developing countries), the necessity of a general regulation of external debt appears to be a precondition for resolving the current account problem in economies of transition. In analysing "the German transfer problem", John Maynard Keynes already showed in the 20s that debt transfer becomes negative for economic development and social stability in debtor countries and even harmful because of the negative repercussions in creditor countries alike.

3 A transformation trap?

A "transformation trap" can open up if external pressures to stabilise the exchange rate (in the case of a trade and current account deficit), and internal pressures to stabilise a fragile monetary sector (due to inflationary pressures) coincide. The vicious circle which is set in motion leads from a trade and current account deficit to inflationary pressure and a fiscal deficit of public finance, to a depreciation of the currency and an increase in the inflation rate, to a loss of competitiveness in foreign markets... This combination of harmful effects on the economy and the society together with a delegitimisation of the new political class can slow down the process of transition and therefore prolong the time period in the "tunnel" of transition, whereas at the same time the "light at the end of the tunnel" is dimmer. Expectations in the transition to free market economies and political democracies will be disappointed and possibly cause some kind of social frustration. This, of course, is not in the interest of social forces in CEECs propelling the transition process, but, moreover, it cannot be in the interest of neighbouring Western European countries. Therefore, there is some need for foreign assistance, maybe via foreign direct investment or via other forms of technical and financial assistance, in order to break the logic of the vicious circle.

But it is not certain whether Western European countries react adequately to the Central and Eastern European necessities just explained. There are several options which can be selected by the EU, and not all of them are benign for transition economies. *Firstly*, Western Europe continues its integration process as if there were no need for assisting Eastern and Central European economies and to give them some more space for facilitating the process of integration into the world market. Perhaps not by accident, the Maastricht I treaty of 1991 coincided with the dissolution of the Former Soviet Union. Maastricht I is a powerful push in the direction of further deepening and not of further widening of Western European integration. In the history of the Western European integration process, "deepening" and "widening" of the European Community/European Union always have been two sides of the same coin. This was the case in 1969, when the customs union was completed (deepening) and Great Britain joined the EC (widening). During the first half of the 80s until 1986, the European Community accepted the Mediterranean Countries (Greece, Spain, Portugal) as new members (widening) and ratified the "Single European Act", a big step towards a Common Market and a Monetary Union (deepening).

But it is doubtful whether widening and deepening remain compatible options in the future. The reason is the degree of economic, social and political integration which has already been reached in Western Europe. The next step, the single currency in a currency union, can only be undertaken in so far as member countries fulfil the rigid Maastricht monetary criteria of convergence. It is rather likely that not all member countries of the EU will be able to meet the convergence criteria so that at the time of the introduction of the common European currency (EURO), not all EU-countries will be accepted as full members of the Currency Union. The European Union, therefore, is becoming a region of "variable geography" (Balladur). The integration process is a "multi-track", "multi-speed", "multi-layer" (John Major) and, therefore, rather contradictory development which might generate more tendencies of divergence than those of convergence.

The consequences of this uncertainty concerning the compatibility of widening and deepening for CEECs must not be negative. On the contrary, it could happen that new forms of co-operation, association and integration will emerge creating better possibilities than before for reshaping the European map of the coming international division of labour, due to the dissolution of a single block of EU member countries.

4 Capital Transfers

The *second* option of the EU consists of actively assisting the process of transition by capital transfers. Capital flows from the Western world to CEECs are not very high, particularly in comparison with foreign direct investment flows into other industrialised or developing countries. The figures concerning the German case are rather telling. They show that nine-tenth of German foreign direct investments (FDI) is directed to the industrialised countries, only 6,2 Billion DM have been invested in reform economies (Hungary, Czech and Slovak Republics, Poland and China). This is no more than 2% of total German FDI. This tendency fits into the diminishing commitment of corporations of OECD-countries in developing countries and the increase of mutual FDIs in the world of developed economies. But even in comparison with the developing countries, private transfers of capital to CEECs in the past have been rather low, although in recent years FDIs in Eastern Europe displayed increasing figures. In the second half of the 90s investment forecasts are quite promising. The problem, however, is that a great part of foreign investment is - comparable to the situation in Latin America - directed to formerly state owned and privatised firms, so that they only partially contribute to net capital formation.

TABLE 2. Foreign Direct Investment in Eastern and Central Europe, 1990-2000 (US$ Million)

Country	1994	1995	1990-95	1996-2000
Hungary	1,146	4,400	11,200	12,968
Poland	1,875	2,500	7,148	21,969
Czech Republic	878	2,500	5,666	15,466
Slovakia	187	200	775	2,150
Slovenia	87	150	501	3,052
Albania	53	75	205	583
Bulgaria	105	150	412	1,428
Romania	340	400	933	4,017
Other Balkans	120	100	300	2,210
Baltics	430	400	1,280	1,890
Russia	1,000	2,000	4,400	26,960
Ukraine	91	113	574	1,400
Other CIS	640	800	2,300	5,085
Eastern Europe	4,791	10,475	27,140	63,847
Eastern Europe & former USSR	6,952	13,788	35,694	99,186

Source: Economist Intelligence Unit, Financial Times, March 23, 1996

At the beginning of the transition process, the whole Eastern European region with about 400 million inhabitants attracted just as much FDI as Ireland and Norway, countries with a population of about 5 million people. The distriibution of German FDI with regard to the perspectives of integration of CEECs into the monetary world market is demonstrated by the following table:

TABLE 3. Levels of German Foreign Direct Investments, end of 1994 (DM Million)

Country and region	German FDI	FDIs in Germany	Balance
All countries	319,448	198,819	120,629
EU	153,245	82,916	69,329
Other OECD	126,142	107,880	18,262
CEECs	6,155	2,029	4,126
Russia	126	852	- 726
Poland	612	258	587
Czech Rep.	1,655	172	1,483
Hungary	2,210	85	2,125
Developing Countries	33,905	5,994	27,911

Source: Deutsche Bundesbank

The German case is not unique. Of the roughly Bn DM 320 of FDI, only Bn DM 6 (less than 2%) are directed to CEECs. Obviously, the transition economies are much more attractive in the official political discourse than in the decision-making procedures of private investors. The reasons are quite simple: First, CEECs have to compete with other regions with regard to FDI. In many cases, obviously, the opportunities for profitable investment in other parts of the world are deemed to be more favourable. Secondly, CEECs have to improve their attractiveness and therefore have to solve the problems of deficient infrastructure, administrative incompetence, security questions etc.. Thirdly, monetary stability issues, i.e. a stable exchange rate and positive interest differentials are important. As we have already seen, it is not easy to achieve these goals, especially since they are contradictory under the circumstances of the transition process. Economic policy under these conditions requires the capability of exercising the art of squaring the circle...

In this stalemate situation of insufficient private capital flows and a negative current balance, official flows of capital assistance to Central and Eastern European Countries come in. The amount of official development aid, other official credits and bank assets of OECD-countries in the developing countries, however, is Bn $ 1714.2, in CEEC only Bn $ 189.0 or 11%. PHARE ("Poland and Hungary Action for Restructuring of the Economy") and TACIS ("Technical Assistance to the Community of Independent States") programmes have transferred Bn ECU 3.3 and Bn ECU 1.4 to Central and Eastern Europe and to the Former Soviet Union, respectively.

Having discussed the problems of integration into the world market of commodities and capital, the world labour market has to be taken into consideration. Migration increased remarkably after the opening of the former Eastern Block. But, since immigration control in Western European countries also became tougher, the consequences of migration flows in the West are not as serious as was predicted five years ago. Nevertheless, the consequences of a brain drain in the CEECs can be negative for the establishment of conditions of diversified quality production and rather expensive for the society as a whole, because the costs of human capital formation have to be covered without the expectations of benefitting from the positive effects on competitiveness.

5 Conclusion

Which conclusion has to be drawn from the experiences with transition to market economies and integration into the already existing international division of labour? Of course, the measures concerned are specific to each country. There is no single rule for transition and integration. But meanwhile, there are patterns of transition policy which can provide some guidelines which have to be established in the specific countries:

First, monetary stabilization is a necessity which dominates all other policy measures. It is an expression of the "hard budgetary constraint" of a monetary market economy.

Second, fiscal discipline is therefore required. But this requirement is not easily fulfilled, for the process of transition can only be performed successfully as long as the legitimisation of the political decision-making entities and the basis of social consent is not undermined.

Third, the exchange rate must be stabilized, either by anchoring the currency to a strong currency abroad or to a basket of other currencies, or by controlling money supply rigidly. Each one of these policy measures means a renunciation of political sovereignty and consequently subjugation to the market forces.

Fourth, industrial policy is required in order to create the material and institutional conditions of systemic competitiveness. Without meso-economic improvements (qualification of labour, research networks, etc.), stabiliziation policies will not work in the long run.

Fifth, economic reforms have to be embedded into political and social reforms. Without the creation of social consent and political legitimisation by participation of the people, economic reforms cannot be undertaken.

Sixth. Not withstanding the importance of these internal measures, emphasis must be placed on the necessity that the integration of CEECs into the world market and into the world community requires changes in the functional and institutional frame work of the world system in general and of the European economic and political order especially.

THE DEVELOPMENT OF SECURITY AND ECONOMY IN THE BALTIC SEA REGION

DIPL.-ING. DIETER HANEL

1. The Change in the Security and Economic Situation in the Baltic Sea Region

During the East-West conflict in Europe, the Baltic Sea region was a region where confrontation was manifested visibly by a high concentration of military and arms as well as by delimitation of the region through the Iron Curtain.

After the end of the Cold War and the disintegration of the Soviet Union and the Warsaw Pact, a complete change in the situation relating to security and the economy took place in the Baltic Sea region. This was due to a change in military policy and to the opening up to free enterprise. Europe is now expanding into the Baltic Sea region, a region in which approximately 50 million people live.

In my paper, I wish to outline this development of security and the economy, devoting particular attention to the following aspects:

- Eastward enlargement of NATO
- New military structures and equipment
- UNO and OSCE
- The role of Russia
- Northeast enlargement of the European Union
- Economic development of the Baltic Sea region
- Armaments industry and conversion

The situation relating to security and the economy in the Baltic Sea region can be depicted in a simplified form as a planetary gear, with the United Nations and the OSCE organisations forming the housing. NATO, EU and Partnership for Peace are the cog wheels with the ten border states.

K. Prunskienė and E. Altvater (eds.), Transformation, Co-operation, and Conversion, 129-140.

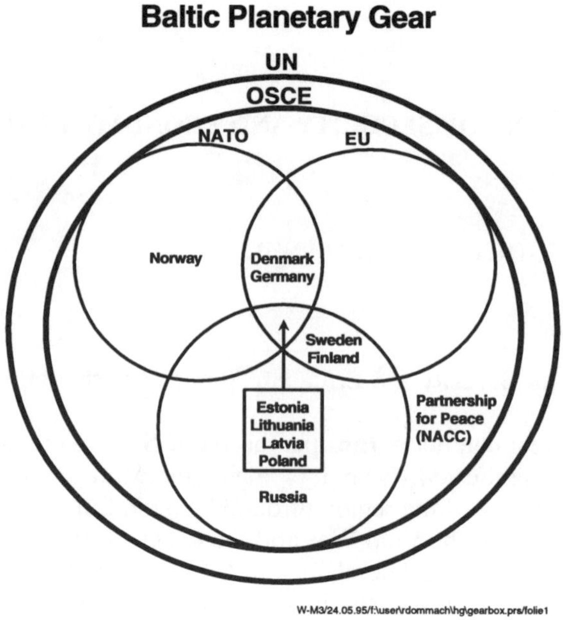

Fig.1 The Baltic Planetary Gear

2. Eastward Enlargement of NATO

After the cessation of the East-West conflict, NATO underwent a thorough change in its functions and structures. Today, it offers its members protection, it is an instrument used for taking measures to secure the peace and an instrument of international) crisis management. NATO still remains the essential basis for European security and is the connecting transatlantic element. NATO's answer to the collapse of the Warsaw Pact is the programme "Partnership for Peace", which has as its aim an extended political and military cooperation with the countries of Eastern and Northern Europe. The partners, however, are not given any guarantees for security.

In addition to this programme, some Central European States are aspiring with differing degrees of intensity to become member states of NATO. Poland has expressed that it hopes to become a member of NATO by the year 2000 and has increased its military cooperation with NATO troops. "Germany and Poland today are already partners in security" said the German Minister of Defence Volker Rühe during a visit made by his Polish counterpart in 1993. The Oder-Neiße cannot form the Eastern boundary line for NATO indefinitely. [1)]

The three Baltic states, Estonia, Latvia, and Lithuania, are in a vacuum where security policy is concerned. We must understand that these states are not satisfied with just a "Partnership for Peace and that they wish to become members of NATO.[2)]

Finland completely accepts the foreign and security policies of Europe, but does not see any need to modify its defence policy. and presently is not aspiring to become a member of NATO.

Sweden, likewise, is not seeking NATO membership, as Prime Minister Carlsson expressed on 17th May, 1995 but wants to decide for itself on its security.[3)]

3. New Military Structures and Equipment

The cessation of the threat in Europe after the ending of the East-West confrontation and the dissolution of the Warsaw Pact on the one hand, and complex global dangers which have made an international crisis management necessary on the other have resulted in an extensive reduction and restructuring of the armed forces. This is accompanied by the withdrawal of Russian troops from Poland, Germany and the Baltic states which was completed in August 1994. The countries of the Baltic Sea region have considerably changed the numbers of their armed forces as the table will show. Around 1 200 000 soldiers are stationed in the Baltic Sea region, approx. 300 000 of which in the Kaliningrad region alone (Table 1).

In March 1995, a new concept for the further development of the German Army[4)] was presented which reflects these changed security conditions and; which plans to reduce the numbers in the armed forces to 340 000. For the Baltic Sea region, this means that a military district command will be formed from the Federal German States of Schleswig-Holstein, Hamburg and Mecklenburg-Vorpommern. The troops in Schleswig-Holstein, in particular the 14th Armoured Tank Division, will be assigned to Neubrandenburg. Then 56 900 soldiers will be stationed in Schleswig-Holstein and Mecklenburg-Vorpommern.

Poland's armed forces have been undergoing an extensive restructuring process since 1990. When this process is finished, the army will be reduced from 15 to 6 divisions.

In matters of security, the Baltic states cooperate closely with the Scandinavian countries within the framework of the Nordic Council. The Baltic states are members of the North Atlantic Cooperation Council, are in the NATO programme "Partnership for Peace" and are seeking full membership in NATO. The armed forces of the Baltic states which are being built up are an expression of the newly won sovereignty. They have announced that they will set up their own UN peace troops which thus will be in keeping with the tradition of Scandinavian countries being engaged under the flag of the United Nations. The Baltic states have increased their military budget considerably in 1994, but the lack of military equipment still continues to be a considerable problem.

TABLE 1. Military Personnel in the Baltic Region

	1991[1]	**1995[2]**
Denmark	29200	31100
Estonia	2000	3500
Finland	32800	31100
Germany	583000 (90)	339900
- Schleswig-Holstein	62000 (90)	34000[3]
- Mecklenburg-Vorpommern		22900[3]
Latvia	2550	6950
Lithuania	7000	8900
Norway	32700	30000
Poland	296500	278600
Russia	2720000 (93)	1520000
Sweden	60500	64000

Source: *1) The Militarx Balance 1992-1993*
2) The Military Balance 1995-1996
3) Der Bundesminister der Verteidigung, Bonn, 15. März 1995

The former Lithuanian Minister of Defence, Mr. Butkevičius, has already been striving for close cooperation between the Lithuanian and German armed forces in 1993[5]. Lithuania intends to build up her armed forces to 20 000, including a rapid reaction troop of 5000.

We must take interest in developing the Baltic Sea region into an area of stability and welfare. The demilitarisation of the area around Kaliningrad is an important part of this task.[6]

4. The United Nations and the Organisation for Security and Cooperation in Europe

All ten states bordering the Baltic Sea are members of the United Nations and the Organisation for Security and Cooperation in Europe (OSCE). The security offered by both these organizations for the Baltic Sea region in the case of a conflict may be doubted due to the negative experiences and inability of these organisations to solve the crisis in the Balkans. The CSCE process has contributed essentially to the liberalisation of the states of Central Europe. However, the OSCE system still does not provide any basis for a collective European security system. The OSCE could in the future be an important instrument for overcoming the lines of demarcation in the eastward expansion of NATO and the EU.[7)]

The results of the OSCE summit in Budapest in December 1994 will contribute "to the stabilization and strengthening of democratic tendencies in Europe, including the Baltic Sea region, as explained in the communiqué of the Ministerial Session of the Council of the Baltic Sea States in Gdansk on May 18-19, 1995.

5. The Role of Russia

Following the disintegration of the Soviet Union, Russia is still going through a politically and economically critical period of transition where long-term perspectives cannot be forecasted.

The war in Chechenia, the breach of the convention for the limitation of conventional arms of 1990, the vehement reactions to the planned eastward enlargement of NATO accompanied by the threat of a world war have all led to the isolation of Russia.

The fact that Russian military policy which has been valid as of November 1993 has not yet been made public is not exactly a step which inspires confidence.[8)]

The politics of Russia have become extremely incalculable. Russia cannot become a member of NATO due to its politically instable situation. Despite this fact, we must still try and integrate Russian interests in a European plan for security. A firm concept for this must be worked out urgently.

An important step in this direction is the joining of Russia to the NATO-program "Partnership for Peace" on May 31st 1995.

6. Northeast Enlargement of the European Union

Sweden and Finland joined the EU on 1st January, 1995. This, in conjunction with the Association Agreement with Poland and the Baltic states has shifted the emphasis within the community considerably to the north. What matters now is to transform Estonias', Latvias' and Lithuanias' prospects of joining the EU into concrete steps which are clearly defined from the point of view of economy and time. In this respect, the European Union has developed a strategy for the integration of these states and will present it in a White Paper with a catalogue of measures at the next summit meeting in Cannes in June 1995. [9)]

It is expected that the Association Agreement between the three Baltic states and the European Union will be concluded in the near future. The next step, which is supported by Sweden and Germany in particular, would be their joining the EU.

The enlargement to the north of the EU also brings with it a new dimension to its foreign and security policy as it includes an important geopolitical region of Europe.

The European Union has defined its aims with regard to a common foreign and security policy in the Treaty of Maastricht. Europe is requested to build up instruments for defence supplementary to NATO. These instruments are to enable the EU to realize its own security interests. The WEU has a military function of regional European nature. The functions defined in the Petersberger Erklärung (Bonn) are humanitarian and peace-keeping missions as well as combat missions within the framework of emergency situations.

Finland and Sweden are, for the time being, content with having an observer status in the WEU.

However, the development of a successful security policy in Europe will only succeed if the new member states and those seeking admission are prepared to play a suitable role in the WEU.

7. The Economic Development of the Baltic Sea Region

Economic recovery is an essential foundation stone for ensuring security in the Baltic Sea region. Economic questions must be of prime importance for the stability of Central European countries.

In 1991, the economic block which existed under Soviet leadership, COMECON, was dissolved. Since then, most of the states have been striving, with various degrees of openness to reform, to establish the structures of free enterprise and to become members of the European Union. Their aim must be to stabilize and privatize an economy which is more or less in ruins and which has a technically outdated industry, an unproductive agricultural economy and an underdeveloped services sector. In the Baltic Sea region, there is an extraordinary market potential for investments and consumer goods.

The commitment of foreign investors is essential for this aim. This commitment, however, is made more difficult by political insecurity and high debts of the companies and burocratic obstacles.

In 1994 the economic performance of the four EU countries, Denmark, Finland, Germany and Sweden was 2303 billion US$, a third of the EU performance.

In 1994, world trade increased in real terms by around 8 %. The upwards trend was particularly noticeable in the Baltic Sea region in Denmark, Finland and Norway. Economic development in Eastern Europe is undergoing a divergent trend. A positive development in Poland is contrasted by a very negative development in Russia. The following chart shows the great difference in the GNP per capita between the five western countries and those of the former Warsaw Pact.

Fig. 2

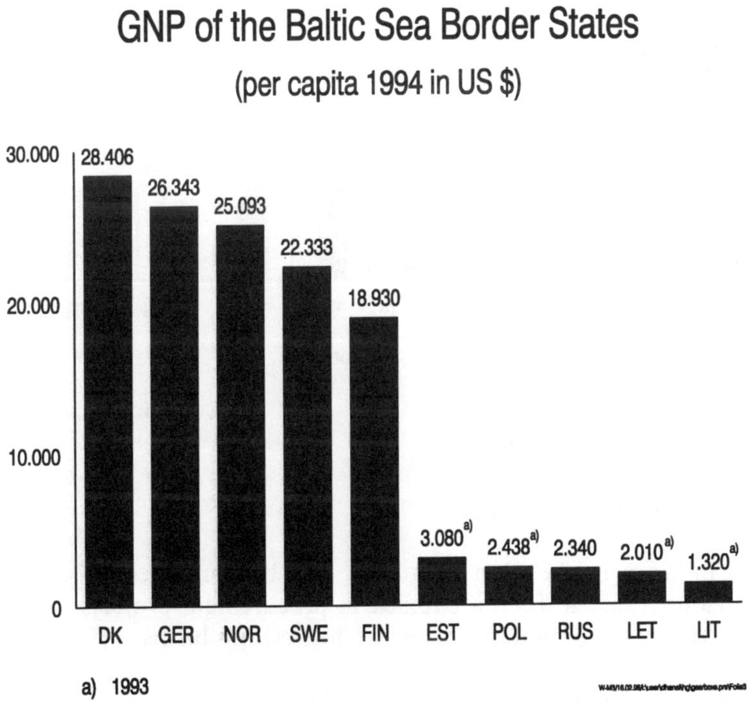

The opening of the borders to the east following the disintegration of the Iron Curtain, and the northeast expansion of the European Union through the admission of Sweden and Finland have led to the intensifying of economic relations between the states of the Baltic Sea region. This has created new markets which are quicker and easier to reach than beforehand and now make an improved network system for land, sea and air indispensable. Sea transport passing through the 70 ports of the Baltic Sea region will increase considerably.

The ten border states of the Baltic Sea complete a considerable part of their economic performances among themselves.

Fig. 3

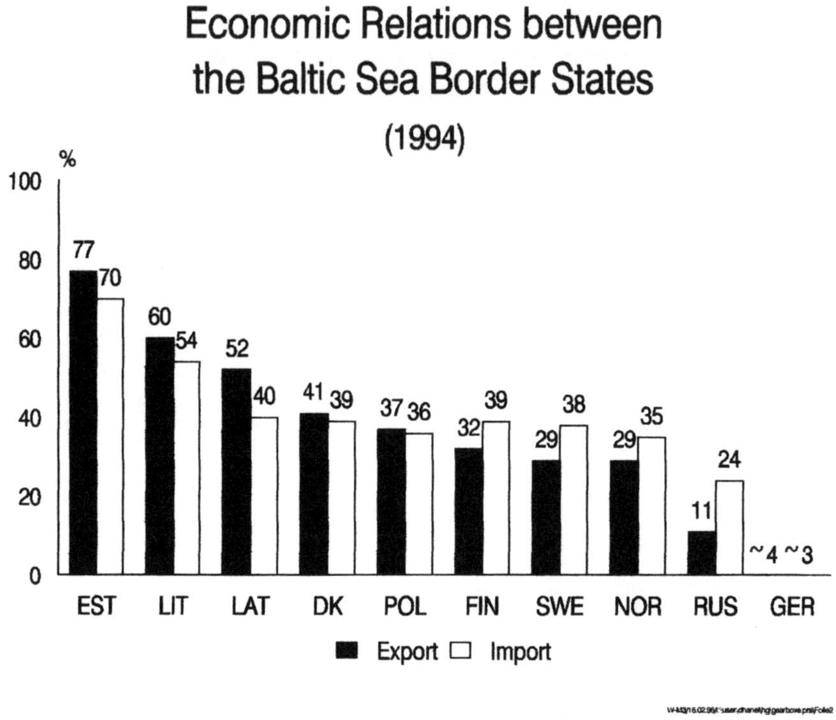

8. Armaments Industry and Conversion

The armaments industry in the East and West has been marked by a drastic reductions in personnel due to both a cutback in procurement projects following the changed functions of the armed forces and to financial problems.

Sweden has an efficient armaments industry which has developed from the concept of neutrality of this country. In addition to covering the needs of the country, this industry is very oriented towards export and recently has turned especially towards the European market. The economic conditions are supported by a "more predominant European dimension in Sweden's foreign and security policy". The armaments industry in Sweden employs approx. 25 000 persons with an annual turnover of 3 billion DM. In the past few years, around 40 % of this industry has been exported. [10)]

There are around 140 000 persons (1990: 280 000) employed in the armaments industry in **Poland**. This relatively powerful armaments industry is oriented to a great degree towards export. However, it has lost the important markets of the former Soviet Union and Iraq.

In **Germany**, the armaments industry has been reduced from 280 000 to 140 000 employees, with a further cutback to 100 000 forecast. Many companies have reached a critical size for the survival of the company. It is now a matter of keeping key technologies which are necessary for the development and production of equipment for the German armed forces and of continuing participation in international cooperation and discussions. In Schleswig-Holstein and Mecklenburg-Vorpommern, those Federal States in Germany which border on the Baltic Sea, around 20000 persons are employed in the defence industry.[11)]

Norway has an efficient armaments industry which concentrates on components. The country largely depends on shipments from abroad. The USA, Sweden and Germany are her main suppliers.

Finland has a small, but efficient armaments industry.

Denmark has a relatively small armaments industry which concentrates on components and on improving the performances of defence equipment.

Lithuania does not have an armaments industry any more.

Estonia has manufacturing plants which are suitable for producing ammunition.

Latvia has no defence industry.

Russia has an important centre for the armaments industry in the St. Petersburg region. In St. Petersburg, 185 out of 476 companies are involved in the armaments industry.

An armaments industry will therefore also be required in the future in the countries around the Baltic Sea. This industry will develop and produce the equipment for the armed forces so that they can fulfil their military tasks whether for national defence or allied defence or within the framework of international crisis management. Capacities which are no longer required will be reduced or will have to be converted to civilian production so that they can survive.

With regard to the conversion of armaments companies, it generally makes little sense from an economic point of view to enter product and market segments which are already firmly established. The high investment costs involved, mostly resulting from extensive employee training, changed company structures, poor customer acceptance and not much market experience make international success rather improbable. For that reason, conversion, or rather diversification, is therefore based on new markets for new products.

The development of new products for new markets means, in economic and engineering terms, a major challenge for a company. The competition is not going to wait until the armaments industry starts taking away their market shares. This does not only apply to the armaments industry. We have seen dramatic examples of this in companies of the former GDR in spite of massive financial support.

The Baltic Sea Border States expend also after the mission of the East-West-Conflict a considerable amount of their budget for defence due to the multidimensional global risks.

Fig. 4

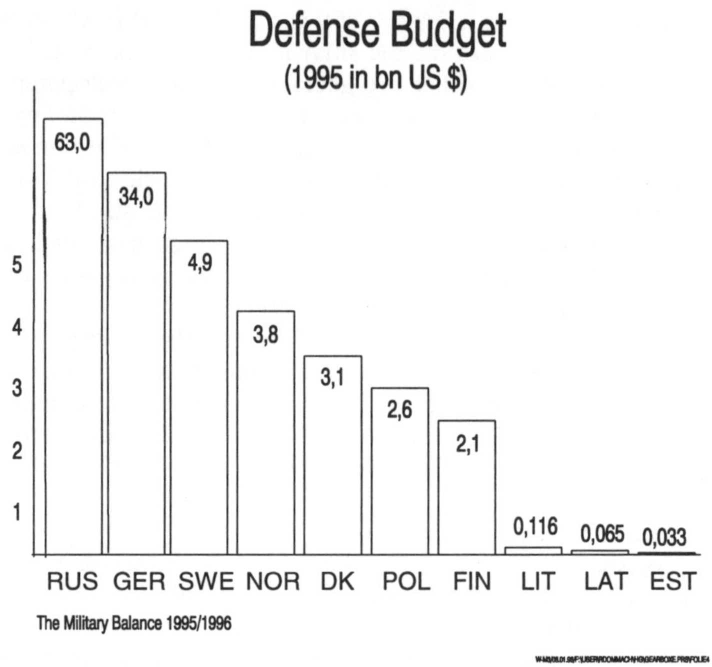

Successful diversification is not possible without earnings from armaments and without technological know-how.

Often, the acquisition of civilian companies or foundation of joint ventures points the way to-success. These procedures, however, have never resulted in new jobs for employees in the armaments industry. Such measures usually do no more than stabilize a company's economic situation.

The economic situation, the ability of the market to absorb new products, the competitiveness and the financial strength of the armaments companies all play major roles in the success of conversion efforts.

Conversion efforts supported by reasonable amounts of state aid could help ensure the diversification of armaments companies.

On the other hand, of course, job market, regional and structural policies cannot be financed by defence budgets.

In view of all these factors, diversification is left up to the companies themselves. This will only work if enough can be earned in armaments and competent planning on the part of the armed forces provides į long-term secure operational basis for the armaments industry.

9. Outline of a Programme for the Baltic Sea Region

1. The Baltic Sea region must grow together so that it can have an increased say in the development of Europe.

2. The aim of the foreign and security policy of all states in the Baltic Sea region must be to make the Baltic Sea a "sea of peace". One way would be a resolute multi-lateralisation of the foreign, security and defence policies of all Baltic Sea states. The "Council of Baltic Sea Border States" would be an essential basis for this.

3. The eastward enlargement of NATO, in conjunction with the enlargement of the EU and the WEU, must be designed as part of an all-European security concept.

4. The economic recovery of the states of Eastern Central Europe is an essential factor for the foundation of social peace and security in the Baltic Sea region.

5. Extensive infrastructure measures for transport and telecommunications must be introduced.

6. Russia must be integrated into a European security .and economic framework in a way which corresponds to the interests of Europe and does not exclude Russia.

7. Maintaining the socio-cultural identity of each state is an important basis for shaping the Baltic Sea region in a "Europe of Regions".

References

1) Polen und Deutschland - Partner im neuen Europa,
2) in: Rühe, Volker: Betr.: Bundeswehr, Berlin, Bonn, Herford 1993
3) Balten fordern NATO-Mitgliedschaft, in: Frankfurter Allgemeine Zeitung, 17.12.1993
4) Handelsblatt, 18.05.1995
5) Ressortkonzept zur Anpassung der Streitkräftestrukturen, der Territorialen Wehrverwaltung und der Stationierung, Der Bundesminister der Verteidigung, Stand: 15.03.1995
6) Soviet Withdrawal from the Baltic States, in: Military Technology,09/1993
7) Garton Ash, Timothy, in: Im Namen Europas, p. 594 f

8) Ganser, Helmut, Die OSZE nach dem Gipfel von Budapest, in: Europäische Sicherheit, p.22-24,04/1995
9) Naumann, Klaus, in: Die Bundeswehr in einer Welt im Umbruch, p.45 ff, Berlin
10) Strategie des Beitritts, in: Europa-Forum, Nr. 04/1995
11) Erbe, Jürgen, Die schwedische Verteidigungsindustrie, in: Soldat und Technik, 04/1995
12) Werhrtechnik in Schleswig-Holstein. Technologie und Kompetenz. Arbeitskreis Wehrtechnik der Industrie in Schleswig-Holstein. Studien- und Fördergesellschaft der Schleswig-Holsteinischen Wirtschaft e. V., Rendsburg/ Kiel 1995.

NAME INDEX

SUBJECT INDEX